# Springer Series in Language and Communication   2

Series Editor: W. J. M. Levelt

# Springer Series in Language and Communication
Editor: W. J. M. Levelt

# The Child's Conception of Language

Editors
A. Sinclair   R. J. Jarvella   W. J. M. Levelt

With 9 Figures

Springer-Verlag Berlin Heidelberg New York 1978

Dr. Anne Sinclair
Dr. Robert J. Jarvella
Professor Dr. Willem J. M. Levelt

Max-Planck-Institut für Psycholinguistik, Berg en Dalseweg 79
6522 BC Nijmegen, The Netherlands

Second Printing 1980

ISBN 3-540-09153-X Springer-Verlag Berlin Heidelberg New York
ISBN 0-387-09153-X Springer-Verlag New York Heidelberg Berlin

Offset printing and bookbinding: Brühlsche Universitätsdruckerei, Giessen
2153/3130-54321

# Preface

It is obvious that the growing child manifests an increasing understanding of his language and facility to use it. A major part of the child language literature is concerned with the child's developing linguistic and communicative competence. Scattered evidence also shows, however, that children become progressively more aware of language *as language*. It is interesting to consider in what ways the internal structure and mechanisms of language become more accessible. Little is known about linguistic awareness of this kind, the role it plays, or how it develops.

When the new Projektgruppe für Psycholinguistik of the Max-Planck-Gesellschaft was founded, "the child's conception of language," in analogy to Piaget's "child's conception of the physical world," became one of the research unit's topics of study. As previous work on linguistic awareness was largely amorphous, we first organized a kind of conference workshop with some of those who had worked in the area. The aims of this meeting were to map out the field of study, detail the phenomena of interest, and define major theoretical issues.

The meeting took place just after the creation of the project group, on May 3-7, 1977. The participants were psychologists and linguists who had either published work on metalinguistic issues in child language, or who could be expected to contribute substantially to the discussion. This book is a direct outcome of that conference, though it is not a complete reflection of the papers presented, or of the discussion that took place. As could be expected, much of what occurred in the meetings was tentative and hypothetical. It did not seem very useful to publish the proceedings as a whole. Rather, we decided to take a somewhat different approach.

Some of the contributions prepared for the meeting appear here only slightly revised. The authors of other papers, whom we felt had also dealt with significant issues, were asked to expand or reformulate them. And in a few cases, other participants were asked to produce a reflection of comments or views they had expressed during the meeting itself. Conference papers by S. Felix, W. Kaper, J. MacNamara, T. Moore, A. Sinclair, C. Snow, and B. Tervoort also deserve special acknowledgement, although they do not appear here in written form.

Our hope was to create a book which would inspire and interest not only those directly involved in the field, but the many psychologists, linguists, anthropologists, and teachers who have a stake in the exploration of linguistic awareness and thinking,and its development. There are many contributions still to be made on this subject, and we hope that this book will serve to stimulate further inquiry by all concerned.

Both the conference and the publication of this volume were made possible by a grant from the Stiftung Volkswagenwerk. Special thanks are due to Mrs. U. de Pagter for helping organize the original conference, and above all to Mrs. H. Kleine Schaars for her care and skill in typing the camera-ready text.

Nijmegen, August 1978                          A. Sinclair
                                               R.J. Jarvella
                                               W.J.M. Levelt

# Contents

*Appendix*

The Role of Dialogue in Language Acquisition

# List of Contributors

BRUNER, J.S., Wolfson College, Oxford University, Oxford, England

CLARK, EVE V., Department of Linguistics, Stanford University,
Stanford, CA 94305, USA

FLORES D'ARCAIS, G.B., Psychology Laboratory, Leiden University,
Leiden and Max-Planck-Gesellschaft, Nijmegen, The Netherlands

GLEITMAN, LILA R., Graduate School of Education, University of Pennsylvania,
PA 19104, USA

GLEITMAN, HENRY, Graduate School of Education, University of Pennsylvania,
PA 19104, USA

HEESCHEN, VOLKER, Ruhr-Universität Bochum, 4030 Bochum, Fed. Rep. of Germany

HIRSH-PASEK, KATHY, Graduate School of Education, University of Pennsylvania,
PA 19104, USA

LUNDBERG, INGVAR, Institute of Psychology, University of Umeå, Umeå, Sweden

MARSHALL, JOHN C., Interfaculty Group on Speech and Language Behavior,
University of Nijmegen, The Netherlands

MORTON, JOHN, M.R.C. Applied Psychology Unit, Cambridge, England, and
Max-Planck-Gesellschaft, Nijmegen, The Netherlands

BERTHOUD-PAPANDROPOULOU, IOANNA, University of Geneva, F.P.S.E.,
Geneva, Switzerland

READ, CHARLES, Department of English, University of Wisconsin,
Madison, WI 53706, USA

SEUREN, P., Philosophical Institute, University of Nijmegen, The Netherlands

SINCLAIR, H., University of Geneva, F.P.S.E., Geneva, Switzerland

SLOBIN, DAN I., Department of Psychology, University of California,
Berkeley, CA 94720, USA

# Introduction

## Causes and Functions of Linguistic Awareness in Language Acquisition: Some Introductory Remarks

W.J.M. Levelt, A. Sinclair, and R.J. Jarvella

Max-Planck-Gesellschaft, Projektgruppe für Psycholinguistik
Berg en Dalseweg 79, Nijmegen, The Netherlands

In order to introduce this book to the reader, we will give a short outline
sketch of the main theoretical and empirical questions that we feel must be
discussed if one endeavors to study the growth of linguistic awareness in
the child. This sketch will at the same time be used to introduce the different papers in this book, seen in the light of the general framework which
emerges. This restriction in viewpoint should be kept in mind. The introduction is not meant to be a review of the contributions to the present
volume, and will not do justice to all (nor perhaps even the greater part)
of the issues taken up in the various papers. It will be no more than an
independent introductory discussion of the field, in which special reference
is made to the work printed here.

The way we will proceed is as follows. Firstly some examples of the
kind of data and phenomena with which we are concerned will be presented.
After this purely descriptive section, the main theoretical issues as we see
them will be discussed in an abstract and rather summary fashion. Few details will be given, since many of these issues are more extensively discussed in different places in the book. Rather, we will try to provide a
peg for connecting some of the theoretical questions raised in the different
papers. This will be done by concentrating on three central issues: the epistemological status of linguistic awareness, its causation, and its possible
functions in language acquisition.

*DATA AND PHENOMENA*

There are different ways of classifying phenomena of linguistic awareness observable in the child. Ultimately, the classification arrived at
will depend on one's theory of structure and function in the child's growing
conception of language. But on a preliminary, atheoretical level, there are
two obvious ways of structuring the phenomena in question. The first of these
is to use a *criterion of explicitness*. Some metalinguistic phenomena are at
the border of awareness, whereas others are clearly the result of very explicit reflection on language. Examples of the first extreme are for instance self-corrections, which occur frequently in normal speech, and which
are readily observable in child speech as well. Re-starts can show that the
child was aware that what he began or was about to say was inappropriate or
incorrect. These phenomena of awareness are very ethereal, going mostly
unnoticed by the listener. For the child speaker they may also pass by on
the stream of consciousness without leaving a trace.

Explicit reflections on language can likewise be observed in children.
One passage from Gleitman, Gleitman, and Shipley (1972) suffices to show this:

> Adult: How about this: *Claire loves Claire.*
> Child (7): *Claire loves herself* sounds much better.
> Adult: Would you ever say *Claire loves Claire?*
> Child: Well, if there is somebody Claire knows named Claire. I
> know somebody named Claire, and maybe I'm named Claire. etc.
> (p. 150)

The ultimate form of explicitness is what linguists regale us with when they
formulate their intuitions concerning the structure or grammaticality of
sentences in the form of rules; and in every child there is a budding linguist.

Between the two extremes one can find various degrees or levels of explicitness. Children not only correct themselves, but also others. This
requires not only awareness of linguistic trouble but at the same time some
ability to formulate what went wrong. Children spontaneously play with
language (cf. Weir, 1962), just as they play with other things. In these
games language becomes divorced from conventional use--it is no longer a
means of communication, but an object of conscious activity. Games involving
rhyme and word substitution can be observed quite early in children (Kaper,
1959). Read's chapter in this volume reports, among other things, on ex-

periments where children are induced to choose words that rhyme with other words. The results provide substantial insight into the children's conception of the English vowel system. Even more explicit manipulation of language can be observed when the child becomes familiar with the printed word. Again, Read's paper gives abundant information about the increasing analytic and syntactic capacities which accompany the acquisition of written language. Lundberg, in his contribution, gives various experimental indications about how these faculties develop. Several linguistic notions in the adult language user seem to derive from early stages of explicit phonological and morphological analysis through which the child has to pass to acquire reading and writing. This seems to be specially true for the notion of "word" itself, to which we will return. The most explicit expressions of awareness are obtained by direct questioning, as in the example given above. One can ask the child about grammaticality, structure, cohesion of sentence parts, case roles (see Braine & Wells, 1978), or appropriateness of speech acts (see especially Robinson & Robinson, 1977), and the older the child the more detailed are the answers one receives. Examples of such explicit question-answering can be found at various places in this book. Flores d'Arcais asked children to judge the similarity of meaning between various connectives and between sentences containing the connectives. Question-answering about grammaticality is reported in Slobin's paper. And Hirsch, Gleitman, and Gleitman interrogated children about ambiguity.

This is not the place to give a full review of the levels of explicitness, and types of phenomena that correspond to them. That is done in a detailed fashion in E. Clark's review paper in this book. Rather we would like to juxtapose this classification scheme with a second one, which is simply the *criterion of developmental stage*. One may ask which forms of linguistic awareness are first to appear, and which come later. Among the first phenomena one can observe is the child coping with the failures of speech acts. If the one-and-a-half year old child addresses its mother verbally, and the mother pretends not to understand, by saying *mhm?* for example, the child is likely to restate the utterance with minor variations, and this may be repeated several times (see for instance, Foppa 1978, and the work of Käsermann referred to there). The variations are probably not random (cf. Marshall & Morton's paper), and seem to indicate that the child is aware that its utterance is in need of some kind of change or improvement. Spontaneous self-corrections, i.e., ones not instigated by an adult, occur almost as

4

early.  These initial expressions of linguistic awareness all seem to serve the function of helping establish effective communication, but it is not much later that verbal play appears; and this seems to imply awareness of a very different form.  Verbal play may involve almost any aspect of language, witness Weir's (1962) surprising observations, and does not serve any immediate communicative function.

It is not much before the age of five or six that the child begins to draw explicit distinctions between the form and meaning of a word or utterance (Papandropoulou & H. Sinclair, 1974, and especially Papandropoulou's contribution to the present volume), and it is not yet known when meanings and functions of utterances become distinguished as occurs in indirect speech acts.  At about the same age, the child develops sensitivity to puns and ambiguity--the separation of form and meaning permitting him to understand and make verbal jokes--and this is evidently an enduring source of enjoyment at school age.  This sensitivity is the main topic of the chapter by Hirsch, Gleitman, and Gleitman.  In a controlled experiment, they show that awareness of ambiguity develops relatively late in childhood.

The acquisition of reading and writing seems to prefigure the adult ability to explicitly discuss the form and function of language.  Not only are the analytic capacities of the child enormously sharpened by the acquisition of spelling and writing, but the visual mode dispels with one of the major problems the child has previously had to cope with in reflecting on his own language: a written word or text remains present and can be studied repeatedly; it need not be held in memory to be reflected upon.  The chapters by Read and Lundberg detail some of these developments.  In addition, however, formal teaching in school provides children with an extensive metalinguistic vocabulary.  Not only does the meaning of the word *word* come into full reach of the child, but so do various technical linguistic terms such as *syllable, noun, verb, subject, object,* etc.  On the other hand, the development of a metalinguistic vocabulary doesn't require explicit teaching.  Slobin, in his paper, describes the early metalinguistic vocabulary of a bilingual child, and Heeschen's article describes a rich metalinguistic vocabulary of an illiterate society still in the stone age.

There is an unmistakable connection between the criteria of developmental stage and explicitness: the older the child, the greater his facility to reflect upon language.  It should, however, be stressed that different levels of explicitness are simultaneously observable at all stages of development.

Self-corrections, for instance, are not only very early expressions of linguistic awareness, but very permanent ones as well: they can be observed in adult language usage and the young child's speech equally well. The same can be said about verbal play, and many other expressions of linguistic awareness. The most comprehensive treatment of relevant data and phenomena in this volume can be found in E. Clark, whose contribution pulls together most of the evidence on linguistic awareness scattered throughout the literature on child language.

## THEORETICAL ISSUES

Though the first priority on entering this still largely unexplored area is to obtain descriptive evidence about the child's growing linguistic awareness, as much theoretical framework as possible should be kept in mind while gathering it. It seems to us that three major theoretical issues should be distinguished. These concern: (1) the epistemological status of linguistic awareness; (2) the causes of linguistic awareness and insight; and (3) the functions of linguistic awareness in development and language use. Let us shortly review these issues in turn.

### Epistemological Status of Linguistic Awareness

What does it mean to be aware of a linguistic state of affairs? Even if one acknowledges that phenomena of linguistic awareness vary greatly, not only with respect to explicitness, but also with respect to content, this very general preliminary question must nevertheless be asked. One answer, which has been rather dominant in the linguistic literature, is that awareness is implicit knowledge that has become explicit. Clearly, this notion goes back to Chomsky's theory of linguistic competence. Competence is tacit knowledge of the language--it exists in the form of linguistic intuitions, which can sometimes be made explicit through questioning or by means of other procedures, and may take the outward form of linguistic judgments (concerning the acceptability of utterances, etc.). Extensive discussion of the epistemological status of this theory can be found in Levelt (1972, 1974), and there is no need to repeat it here. The main conclusions, however, are first that if there is such a thing as a unified competence underlying all linguistic behavior, then explicit intuitions have at most a highly indirect and involved relation to this base of tacit knowledge. Contrary to what has generally been claimed, the relations between explicit intuitions and underlying competence are *less* direct than those between phenomena of primary language usage (speaking, listening) and competence. Second, linguistic judgment is a form of behavior which should be explained in its own right, just like any other form of linguistic behavior. It has no special epistemological status. This is, more generally, true for linguistic awareness: it is a phenomenon to be explained. There is no reason whatsoever to assume that it has a special "hot line" to implicit knowledge. This point is taken up further in Seuren's contribution to this book. That phenomena of linguistic awareness have no privileged status, and require explanation just as other linguistic phenomena do, is again strongly argued by Read, and by Hirsch et al. in this volume.

These statements should not be read as implying that linguistic awareness is a "mere" epiphenomenon which accompanies certain unconscious linguistic processes but never play a functional role in them itself. At this point one should make an important epistemological distinction. If one is aware of a linguistic entity X, X might be one of the *causes* of this awareness. For instance, X could be built-in property of the language-producing mechanism, and be part of the cause of a speaker expressing some form of awareness of X. On the other hand, X could be nothing else than the *intentional object* of the user's awareness, without having any real existence outside this intention. This is what could be called the "epiphenomenon" situation. Finally, X could be *both* cause *and* intentional object: the thing one is aware of does have a real existence, and (partly) caused awareness to come about. This final view is very much in line with a realist interpretation of perception. (The real existence of object O in part causes the perception of O.) A similar realist interpretation of "being aware of linguistic entity X" seems rather strong, and there is reason to be careful not to jump too rapidly from an intentional object interpretation of linguistic awareness to a cause interpretation.

Perhaps the mechanism proposed in Marshall and Morton's paper comes closest to what we have in mind here. We will return to this matter in the next paragraphs. For the moment, let us simply stress that neither the adult, nor the child, can become aware of the biological or mental machinery involved in language use. In history a similar point has been made over and over again with respect to other cognitive abilities. It is what Bühler (1907, 1908) claimed for conscious phenomena in thinking, Piaget (1974a, 1974b) for complex motoric and cognitive performance, and Nisbett and Decamp Wilson (1977) for a large variety of cognitive processes. H. Sinclair in this volume discusses the Piagetian point of view. She explains that "becoming aware" is a mental activity which interacts with other cognitive processes. The subject is almost fully unconscious of the internal mechanisms of cognitive functioning and of those involved in "becoming aware." Seuren, in his paper, makes the same point for linguistic functioning. Nisbett and Decamp Wilson's main point is that there is very little evidence in the literature that people *are* conscious of many of their own mental processes. Awareness seems to be restricted to the outcome or results of such processes, and if people do report on processes, this is--they contend--usually a logical reconstruction of how such a result might have come about (often in the form of a motivation) rather than a memory trace of the process itself. It should be added, however, that the distinction between process and outcome is not so obvious as it may seem at first sight. If knowledge is procedural, as is often claimed in the artificial intelligence literature, then results of procedures are procedures again, and the distinction made by Nisbett and Decamp Wilson fades considerably.

It is especially in the case of "slow" mental processes, such as thinking in chess (De Groot, 1965), that subjects seem to be able to describe the process more or less on line. In this way protocol analysis contributed quite essentially to our knowledge of problem solving (Newell & Simon, 1972). Whether this is in agreement with Nisbett and Decamp Wilson, or Bühler for that matter, is an issue in itself; what is relevant here is that, as opposed to problem solvers, language users show practically no insight into the way they perceive or produce language. Protocol analysis is simply not a feasible technique for analyzing language understanding or production. What is involved is, as a rule, very much in the dark for the language user himself. It is thus all the more interesting to investigate under what conditions the language user *does* become aware of his speaking or listening processes.

Read, in his article, shows that linguistic awareness can be provoked in children, or more precisely that by questioning a child about his language, the child acquires access to linguistic structure of which he showed no spontaneous awareness beforehand. It is indeed important to make a distinction between spontaneous manifestations of awareness (in daily usage), and potential accessibility of linguistic processes and structures. There is no doubt that the latter has definite limits, but these limits and how they retreat as the child develops deserve careful examination.

*Causes of Linguistic Awareness*

Skills like driving a car can become almost completely automatic with increasing experience. By "automatic," we mean that the activity requires little conscious attention. For example, while at the wheel an experienced driver can simultaneously and attentively discuss a complicated issue with his fellow passenger. Skills are, by definition, automated activities, in which low-level decisions (such as steering or accelerating) do not require conscious attention. Attention can be reserved for more general, high-level reflection and decisions (such as where to go, and when to get the tank filled). In performing complex skills, moreover, low-level decisions often have to be taken far too rapidly to be handled by conscious decision-making. If they are taken consciously, the activity may become disturbed and break down: indeed, this is what we observe in beginning drivers, who do have to pay attention to nearly all the elements in the total action pattern. Linguistic skills are no different in this respect (see Levelt & Kempen, 1978; Levelt, 1978). In speaking, for instance, low-level decisions—including, for example, choice of syntactic frames and articulatory activities—apparently require no, or almost no, conscious planning decisions. Attention is usually directed to the content of what is going to be expressed, not to the language used for it. The latter develops in a largely automated fashion.

For any skill, however, there are two obvious circumstances where conscious reflection or decision-making, and therefore awareness, occur. The first one is during *skill acquisition*, as in the example of learning to drive. Automatization of elementary activities is often preceded by a stage of conscious learning of these activities. Whether a conscious stage also precedes automatization of elementary speech activities during language acquisition is still an open question. We will return to this point in the next section.

The second circumstance where awareness can be observed in the execution of skills is at moments of *failure*, i.e., when unexpected or undesired results occur. If suddenly the car one is driving starts to skid, one becomes fully aware of what is happening, and conscious decision-making may result (if there is sufficient time left). A similar claim can be found in Piaget (1974a) where the child becomes aware in cases of disequilibrium, i.e., where the automatic regulations in performing intentional acts are no longer sufficient to attain a goal. It is this mechanism of matching aim and result which H. Sinclair in her chapter applies to speech acts. In the same way, a major cause for linguistic awareness could be failure in communication, that is, in speaking or understanding. Repairs made while speaking, or the registering of *what?* while listening, may occur when automatic processing fails to yield the result being sought: a speech error is corrected or there is an attempt to remedy a lapse in understanding. Conscious intervention is then required, and the language user is--momentarily at least--in some fashion aware of the linguistic entity that caused the problem. Problems of this type abound for the young child, of course, precisely because his linguistic skills are still in a very preliminary stage of development. (For a case study of linguistic awareness in a child exposed to several languages, see Slobin's chapter.) In fact, several papers in this book show how essential this notion of failure is for the explanation of early linguistic awareness (cf. especially the chapters by E. Clark, H. Sinclair, and Marshall & Morton). There is no need to elaborate this point here, and we turn to a set of causes which may be involved with more explicit forms of linguistic awareness.

Explicit reflections on language can be obtained from the child (as well as from the adult) by asking explicit questions, such as *Can one say X?* or *Is X good English?* How subjects actually manage to answer such questions is largely unknown, but one plausible procedure would be for the language user to try to think of a situation in which X *can* be said. If it is easy to imagine such a situation, the answer may well be *Yes*, if it is too difficult to think one up, it may be *No*. Levelt et al. (1977) tested this theory by giving subjects verbal material of borderline grammaticality which related to situations that were either easy or difficult to imagine. It was indeed the case that the former material was more frequently judged to be grammatical than the latter. Subjects also answered more rapidly for material which was easy to imagine. Nevertheless, results showed that it was unlikely that the subjects, in conceiving their answers, had to find a "full" interpretation for the verbal item presented. This was concluded from a comparison with reaction times for a paraphrase task using the same material, which had a very different pattern. It is argued in that article that intuitions concerning acceptability may result from rather shallow semantic processing; this is in line with the findings of Mistler-Lachman (1972) on intuitions concerning meaningfulness. In the next section we will return to the possible functions of such shallow processing. Answering questions about acceptability and paraphrase (as in Flores d'Arcais' paper), about morphological structure (as in Read's paper), about word length (as in Papandropoulou's and Lundberg's contributions), and so forth, requires what Lundberg calls an "attention shift," i.e., a shift from content to form. Lundberg shows that the difficulty of making this shift adheres to general Gestalt principles. The stronger the semantic "Gestalt," the harder to break it down.

These few remarks about causation are admittedly vague and of a general nature; for these reasons, their value is clearly limited. The origins of linguistic awareness may in reality be as varied as the phenomena themselves. What induces the child to play linguistically, either alone in self-talk or with others by joking, to play rhyming games, and the like? These questions are far from being answered, but it might nevertheless be worthwhile to ask an even more abstract question: Is the genesis of linguistic awareness universal? And if so, which forms of awareness are universal, and which are language or culture specific? A major distinction needs to be made here between literate and illiterate societies since learning to read and write fosters the development of linguistic notions and reflections that might otherwise not easily occur. The notion of "word," let alone notions such as "syllable," "sentence," and other general illustrations of phonological or syntactic insight, may be very writing-dependent.

Thus, Heeschen in this volume shows that the metalinguistic vocabulary of the Eipo speech community, a neolithic mountain people of New Guinea, is bound by content and appropriateness, and shows little concern for language structure proper. Like other groups, the Eipo are sensitive to various dialect differences with and between their own and neighboring cultures; their linguistic beliefs and taboos require that words be "chosen" and "used" with care. Even so, the conclusion must be guarded, since Heeschen also found that there was high agreement on fixed word and clause orders among the speakers of the language he studied. If, on the other hand, the "failure causes awareness" theory is correct, then self-corrections and other similar phenomena should be observable in all societies, since "skillfulness" is a property of all normal language use. It is much less trivial to predict whether rhyming, or for that matter intuitions of acceptability, are universal. There is a major task here for linguistic anthropologists.

A final remark should be made about asking the child metalinguistic questions, such as *why*-questions, that require whole explanations as answers. Older children can handle such questions better than younger ones, but it is probably false to ascribe this fact solely to a greater reflexive awareness of language. Explaining is a complex verbal activity; the child has to express more or less complex ideas or facts in a proper linguistic form.

This skill is involved in explaining the way or in telling narratives, as well as in giving metalinguistic answers. The development of the child's capacity to explain verbally shouldn't be confused with his growing capacity to reflect on language: methodological care is required to keep these issues apart.

## The Functions of Linguistic Awareness

For what use do we develop linguistic awareness? As Heeschen points out, awareness occurs when there is contact between languages, and historically, rather than being exceptional, such contact is the rule. We also know that a certain sophistication in consciously manipulating language is a requirement for learning to read and write (see especially Read, this volume). Still, it seems unlikely that evolution has endowed us with awareness of language for just these purposes. Such skills seem fairly tangential even to basic language use. The more fundamental functions of linguistic awareness should probably be sought for in the facilitative role it plays in face-to-face communication, and in learning to communicate. Let us consider these in turn.

Linguistic awareness can facilitate communication. In the preceding section it was suggested that awareness can be provoked by failure. It may arise momentarily when a conversation goes awry. Conscious repair may be an effective means of coping with such moments of difficulty and helping prevent a deeper breakdown in communication. This situation is not different from the one for other complex skills. At a moment of failure, i.e., where the automated procedures produce an undesired result, conscious correction can often prevent more serious, subsequent failures. What is especially interesting about conscious linguistic repairs is that they themselves have a rather complex but systematic structure. Scheg̃loff et al. (1977), in a recent paper, describe such regularities, and remark that "An adequate theory of the organization of natural language will need to depict how a natural language handles its intrinsic troubles" (p. 381). Language is a skill which, through its sheer complexity, is very failure-prone, and it has developed its own mechanisms of coping with this, among them conscious repairing. E. Clark's paper in this book especially stresses this function of linguistic awareness.

Other mechanisms may have developed in order to *prevent* failure. In the last section we mentioned the shallow processing which seems involved in judging the acceptability or meaningfulness of an utterance. One might guess that such "monitoring for meaningfulness" does not only take place as a consequence of a direct metalinguistic query, but is a general mechanism involved in normal language understanding. It is known that we do monitor speech we hear for specific acoustic features in order to get some "standards" for the correct interpretation of what is to follow. Broadbent and Ladefoged (1960) showed that a listener checks for speaker-specific vowel formants in the first few syllables a new speaker says, and uses these to interpret the vowels to come. It is not likely that a listener makes only phonetic checks. More plausibly he also performs syntactic and semantic ones and these may also lower the probability of comprehension failures. Presented with Chomsky's *Colorless green ideas sleep furiously* the listener probably will not go into an elaborate semantic memory search in order to understand: a shallow preliminary analysis makes it clear that the sentence is meaningless. In this way, superficial checks and pretests may have a monitoring function which prevents unnecessary processing and false interpretation.

Apart from facilitating communication in various ways, linguistic awareness may play a functional role in the acquisition of communicative skills. This was mentioned in passing in the preceding section in connection with the acquisition of skills in general. The point was made there that, to acquire complex skills, a great deal of attention is often required in order to correctly perform the elementary actions of the complex action pattern. Fluent integration of the skill only follows the automatization of activities at the elementary level. It is not clear, and certainly not obvious, that this is true for language acquisition as well. But if it is, one would expect young, i.e., less skilled, language users to be more aware than mature speakers of some of the details involved in speaking. The younger child might need to work consciously on superficial aspects of its language, such as morphological and articulatory details, whereas the older child is fluent in these respects and can direct his attention elsewhere. The evidence for this somewhat paradoxical hypothesis--that the young child is more aware of certain properties of language and speech than the older one--is admittedly scarce. Read (this volume) does show that some phonetic judgments deteriorate with age. Adults who are familiar with written forms have to work in order to re-acquire distinctions that kindergarten children can make easily. Zei (1978), on the other hand, found no evidence that five-year olds were better able to explain the articulatory events involved in speaking than nine-year olds. In her study, no comparison was made with adults, however. The issue deserves further attention. It may be especially worthwhile to investigate which aspects of early speaking and listening activities do receive explicit attention during acquisition, and which do not. Knowing this might tell us something about the procedures by which the child acquires linguistic skill, and about how speaking and listening are organized at different ages.

This brings us finally to the issue of what feedback structure is involved in the acquisition of speech skills. It is essential in the learning of any skill that the learner acquire internal standards in order to evaluate his own performance. Feedback is not only necessary for comparing standard and performance so that deviations which exceed a certain criterion may be corrected, but also for the development of the internal standards themselves. One theoretically-minded approach, developed by Marshall and Morton in their contribution to this volume, is to describe early forms of linguistic awareness in terms of feedback mechanisms which are involved in the acquisition of basic linguistic skills. The child's monitoring of (others' and especially his own) speech may provide him with the opportunity to check and thereby raise his own standards of production. Marshall and Morton put forth the interesting hypothesis that, as the child learns to speak, the perceptual system "teaches" the production system in approximately this way.

These and similar notions of the functions of linguistic awareness in language acquisition may well have some face validity. However, as yet none of them has been tested experimentally. And experimental tests *are* necessary. Descriptive evidence will not be sufficient, since correlational findings cannot answer the essential question of whether conscious awareness of language does contribute to first language learning, and if so, whether it is a necessary condition as well. Because of the ethics of human subjects research, the latter part of this question will probably never be answered. The former part, however, is certainly within the reach of systematic experimental study. We hope that this book will contribute to the development of such research.

## REFERENCES

Braine, M.D.S., & Wells, R.S. Case-like categories in children: The actor and some related categories. *Cognitive Psychology*, 1978, 10, 100-122.

Broadbent, D.E., & Ladefoged, P. Vowel judgments and adaptation level. *Proceedings of the Royal Society*, 1960, 151, 384-399.

Bühler, K. Tatsachen und Probleme zu einer Psychologie der Denkvorgänge. *Archiv für die Gesamte Psychologie*, I. Über Gedanken, 1907, 9, 297-305; II. Über Gedankenzusammenhänge, 1908, 12, 1-23; III. Über Gedankenerinnerungen, 1908, 12, 24-92.

De Groot, A.K. *Thought and choice in chess*. The Hague: Mouton, 1965.

Foppa, K. Language acquisition--a human ethological problem? *Social Science Information*, 1978, 17, 93-105.

Gleitman, L.R., Gleitman, H., & Shipley, E.F. The emergence of the child as grammarian. *Cognition*, 1972, 1, 137-164.

Kaper, W. *Kindersprachforschung mit Hilfe des Kindes: Einige Erscheinungen der kindlichen Spracherwerbung erläutert im Lichte des vom Kinde gezeigten Interessen für Sprachliches*. Groningen: Wolters, 1959.

Levelt, W.J.M. Some psychological aspects of linguistic data. *Linguistische Berichte*, 1972, 17, 18-30.

Levelt, W.J.M. *Formal grammars in linguistics and psycholinguistics*. Vol. III Psycholinguistic applications. The Hague: Mouton, 1974.

Levelt, W.J.M. Skill theory and language teaching. Fourth Colloquium in Theoretical Models in Applied Linguistics. To appear in *Studies in second language acquisition*. Bloomington: Indiana University Linguistics Club, 1978.

Levelt, W.J.M., Van Gent, J.A.W.M., Haans, A.F.J., & Meyers, A.J.A. Grammaticality, paraphrase and imagery. In: S. Greenbaum (ed.), *Acceptability in language*. The Hague: Mouton, 1977.

Levelt, W.J.M., & Kempen, G. Language. In: J.A. Michon, E.G.J. Eykman & L.F.W. de Klerk (eds.), *Handbook of Psychonomics*. Amsterdam: Elsevier, 1978, in press.

Mistler-Lachman, J.L. Levels of comprehension of normal and ambiguous sentences. *Journal of Verbal Learning and Verbal Behavior*, 1972, 11, 614-623.

Newell, A., & Simon, H.A. *Human problem solving*. Englewood Cliffs: Prentice Hall, 1972.

Nisbett, R.E., & Decamp Wilson, T. Telling more than we know: verbal reports on mental processes. *Psychological Review*, 1977, 84, 231-259.

Papandropoulou, I., & Sinclair, H. What is a word? Experimental study of children's ideas on grammar. *Human Development*, 1974, 17, 241-258.

Piaget, J. *La prise de conscience*. Paris: Presses Universitaires de France, 1974(a).

Piaget, J. *Réussir et comprendre*. Paris: Presses Universitaires de France, 1974(b).

Robinson, E.J., & Robinson, W.P. Development in the understanding of causes of success and failure in verbal communication. *Cognition*, 1977, 5, 363-378.

Schegloff, E.A., Jefferson, G., & Sacks, H. The preference for self-correction in the organization of repair in conversation. *Language*, 1977, 53, 361-382.

Weir, R.H. *Language in the crib*. The Hague: Mouton, 1962.

Zei, B. Psychological reality of phonemes. *Child Language*, 1978, in press.

# Part I

# Empirical Studies

# Awareness of Language: Some Evidence from what Children Say and Do

Eve V. Clark

Stanford University, Stanford, CA 94305, USA

Children begin to reflect on certain properties of language at an early age. They comment on their own growing linguistic abilities--for example, "When I was a little girl I could go 'geek-geek' like that. But now I can go 'This is a chair'," from a child aged 2;10 (Limber, 1973). They reject wrong pronunciations--for example, when a child teased his younger brother by mimicking his pronunciation of *merry-go-round*, "mewwy-go-wound," the younger brother firmly corrected him, "No, you don't say it wight" (Maccoby & Bee, 1965). And they comment on how others, usually younger children, speak--for example, a five-year old, hearing his brother pronounce *spoon*, asked their mother, "Why does he say *coom*?" (Weir, 1966, p. 164).

The study of what children are aware of provides one way of finding out what their conception of language is. But this requires that we first establish what reflective abilities they have, when and how these develop, and what role they play in acquisition itself. In the present paper, I shall review some of the evidence that children are aware of and able to reflect on certain properties of language.

People can be aware of their language at many different levels, from the automatic, virtually unconscious monitoring of their own speech to the rapid switching of languages by professional translators to the detailed analytic work of linguists. The first signs of an ability to reflect upon language begin to appear at about age two. They include:

   (i)   Spontaneous corrections of one's own pronunciations, word forms, word order, and even choice of language in the case of bilinguals;

(ii)   Questions about the right words, the right pronunciation, and the appropriate speech style;

(iii)  Comments on the speech of others: their pronunciation, accent, and the language they speak;

(iv)   Comments on and play with different linguistic units, segmenting words into syllables and sounds, making up etymologies, rhyming, and punning;

(v)    Judgments of linguistic structure and function, deciding what utterances mean, whether they are appropriate or polite, whether they are grammatical;

(vi)   Questions about other languages and about languages in general.

Although a list like this makes the study of children's awareness seem fairly straightforward, the criteria for assessing awareness are not always clearcut.

Somtimes it is difficult, for example, to distinguish implicit judgments about language from everyday use.  Consider two-year-olds who respond to well-formed commands (e.g., *Throw the ball!*) but not to telegraphic ones (*Throw ball!* or *Ball!*).  Discrimination of these two types of command could be the result of awareness to differences in form between them at some level.  But equally, it could be the result of understanding only a single type of command.  Simply using a language is not the same as reflecting upon it.  Older children, of course, are more likely to use explicit judgments about language (e.g., "Only Daddy can say that") and their questions about language are more obviously reflective (e.g., "Is it *ber*-fore or *be*-fore?").

Evidence that children are aware of different facets of their language is scattered throughout the acquisition literature.  What I have tried to do is gather together some of the observations that have been made in order to present a preliminary taxonomy of the ways in which children reveal their growing awareness of language--of its structure, sound system, lexicon, and morphology as well as its function and the social rules for its use.  Once we have looked at these, I will return to the more difficult problem of what it means to be aware and the role that such awareness might play in the actual acquisition of language.

## MAKING JUDGMENTS ABOUT LANGUAGE

Even very young children can make judgments about appropriateness, complexity, and even about possible forms.  In this section, I will briefly take up some of the studies pertinent to such judgments.

## Appropriateness

Judgments of appropriateness have involved explicit choices of which of two utterances would be more polite, choices of which types of speech go with different roles, and adjustments made by speakers to different listeners. Judgments of relative politeness or niceness in making requests were first studied by Bates (1976). Italian children were asked to request sweets from an old woman puppet, and judge which request was more likely to be successful from pairs such as *Voglio un dolce* ('I want a sweet') and *Vorrei un dolce* ('I would like a sweet'). Children as young as four-and-a-half were fairly consistent in choosing the less direct request as the nicer one, while younger children did not make consistent choices.

In another recent study, Andersen (1977) asked children between four and seven to "do the voices" for different puppets, for instance, a father, mother, and baby in one setting, and a doctor, nurse, and child patient in another. She found that even the youngest children adjusted their speech to differentiate among the three family roles, but did less well on doctor-nurse-patient roles and teacher-pupil roles. Appreciation of how one should speak in particular situations may be present for some children even earlier than four. Some of the diary records report children objecting to inappropriate usage. One three-year old gave up pretending to talk to her uncle on the telephone when her mother used the wrong voice: "That isn't the way. I ain't going to talk to you (3x). I ain't going to talk to you any day (3x). 'Cause you don't talk right. I ain't going to telephone to you any day, 'cause you don't talk right " (Brandenburg, 1915, p. 106).

There have also been a number of studies of how speakers adjust their speech when addressing particular listeners. Shatz and Gelman (1973) found that four-year olds consistently used shorter, simpler sentences to two-year olds than to four-year olds or adults (see also Gelman & Shatz, 1977). This kind of adjustment to the age, status, and even sex of the listener is evidently acquired fairly early (see further Berko Gleason, 1973; Sachs & Devin, 1976; Edelsky, 1977; Snow & Ferguson, 1977).

## Complexity

Besides appropriateness, young children also seem capable of judging relative complexity. Shipley, Smith and Gleitman (1969) found that with children between 1;6 and 2;6, those at the one-word stage consistently responded to very simple commands like *Throw ball!* or just *Ball!* but ignored more com-

plex ones like *Throw me the ball!* Children at the two-word stage did just
the reverse: they responded consistently to the longer, adult-like commands
and ignored the one- and two-word versions. This, of course, is not necessarily evidence of awareness since we are in the borderline area of implicit
knowledge inferred from the children's response to instructions.

Somewhat older children, however, are able to make rather similar judgments at one remove from direct responses. Scholl and Ryan (1975) asked
five- and seven-year olds to identify the speaker of different utterances by
pointing to a picture of an adult or of a small child. The children usually assigned structurally more primitive sentences to the younger speaker,
e.g., questions and negatives like *What the cow say?* and *We not go home.*
More complex versions of the same sentences, *What can the cow say?* and *We
cannot go home*, were attributed to the adult speaker.

## Form

Very young children also seem capable of making some judgments about
form in the language they are acquiring. For example, Gleitman, Gleitman,
and Shipley (1972) asked two-year olds to judge which sentences were "silly"
from among a set of telegraphic and full forms, with either normal word order
or noun and verb reversed (e.g., *Bring ball, Bring me the ball, Ball bring,
Ball me the bring*). Two of the three children not only offered judgments but
also volunteered "corrections" of the sentences they found silly (e.g., *Pull
wagon → Pull wagon me, Song me the sing → Sing me the song, Fill the mailbox →
Put mail in the mailbox, Ball me the throw → Throw the ball*). Although the
responses from two-year olds in this task were limited, Gleitman and her
colleagues concluded nonetheless that children under three could in some
sense already "contemplate the structure of language" (p. 147) (see also
de Villiers & de Villiers, 1974; James & Miller, 1973).

Gleitman and her colleagues also interviewed older children between
five and eight and asked them to make judgments about sentences, some mal-
formed and some not. These children were willing to accept a sentence like
*My sister plays golf* and reject *Golf plays my sister*, just as adults would.
And even five-year olds were able to give relevant accounts of what was wrong
with most of the deviant sentences. As an illustration, consider the following exchanges between Gleitman and her daughter Claire, aged seven:

(1)    LG: How 'bout this one: *Boy is at the door.*
       Claire: If his name is boy. You should--the kid is named John,

see? *John is at the door* or *A boy is at the door* or *The boy is at the door* or *He's knocking at the door.* (p. 149)

(2)　　LG: How 'bout this: *Claire loves Claire.*

Claire: *Claire loves herself* sounds much better.

LG: Would you ever say *Claire loves Claire?*

Claire: Well, if there's somebody Claire knows named Claire. I know somebody named Claire and maybe I'm named Claire.

LG: And then you wouldn't say *Claire loves herself?*

Claire: No, because if it was another person named Claire--like if it was me and that other Claire I know, and somebody wanted to say that I loved that other Claire they'd say *Claire loves Claire.* (p. 150)

The studies of Gleitman et al. make it very clear that by the age of about five, some children are capable of reflecting on their language in a very sophisticated way (but see Moore, 1975; Hakes, Evans, & Tunmer, in press).

Children of this age can also make some judgments about morphology. Bogoyavlenskiy (1957/1973) asked Russian children about differences in meaning conferred by the addition of diminutive, augmentative, and agentive suffixes to nonsense words used in short stories. The word *lar* (applied to a make-believe animal) was distinguished from *larishche* (a big lar) and from *larёnok* (a small one). However, very few of the children (aged between five and eight) were able to identify the word *endings* as the elements that actually made the difference to the stem meanings.

Lastly, children will make explicit judgments about the sound system, usually regarding their own ability to pronounce a specific word. N. Smith (1973) reported the following exchange between himself and his son, aged two-and-a-half:

Father: Say *jump.*

Child: Dup.

Father: No, *jump.*

Child: Dup.

Father: No. *Jummmp.*

Child: Only Daddy can say *dup!* (p. 10)

Smith's son also commented spontaneously that he could now pronounce some new word or sound sequence within a word. After nearly a year of pronouncing *quick* as "kip," he announced: "Daddy, I can say *quick*" (p. 10). These observations are paralleled in Leopold (1949), whose older daughter Hildegard

showed a similar interest in her own improved pronunciation. She had at
first pronounced *merry-go-round* as [mɛkəriraʊnd], but at the age of 4;1,
announced mastery of the right pronunciation: "Watch my mouth: merry-go-round"
(vol. 4, p. 57). In speaking German, she had had considerable difficulty with
the pronunciation of *Verzeihung*, confusing it with *Zeitung*, but around age
four said one day: "Look at my lips: Verzeihung ([fər'saiʊŋ])." A few months
later, when talking about an upcoming visit to her German grandfather, Hildegard commented on the difference between the English and German pronunciations
of her name: "Opa might call me Hildegard [-d], in German though: Hildegard
[-t]," (vol.4, p. 75).

Judgments like these--of appropriateness, of complexity, and of form--
all suggest that children become aware of language on several different dimensions at once. They attend to the kind of speech appropriate to different
ages and social roles and at the same time begin to distinguish odd sentences
from acceptable ones, simple sentences from more complex ones, and their own
pronunciations from those of others. Their explicit comments on language
seem to begin around the age of three.

## APPLYING RULES TO NEW INSTANCES IN PRODUCTION

Another indication that children are aware of language is their adher-
ence to rules in producing it. Presented with unfamiliar or nonsense words,
for example, they can add the appropriate plural, possessive, or past tense
endings. Their ability to do this, many have argued, is a measure of their
knowledge of the general rules for applying specific inflections. Berko
(1958), using nonsense words that the children could not have heard before,
was able to show that five- and seven-year olds were very good at applying
appropriate noun and verb endings in English. They had more difficulty with
endings for the comparative (*-er*) and superlative (*-est*).

In a similar study of Russian children, Bogoyavlenskiy (1957/1973) used
the names of various objects and asked five- and six-year olds what they would
call a baby ostrich, a baby oak-tree, a baby nose, etc., to see whether they
could supply the appropriate diminutive endings. All the children were successful although some did not distinguish between those diminutives that are
normally applied only to animate nouns and those that are not so restricted.

Even younger children sometimes give striking evidence that they have
mastered a particular word ending. They over-regularize and apply the ending to words that do not, in fact, take it, producing every English past

tense, for instance, with the suffix *-ed*, e.g., *breaked, goed,* and *doed,* alongside *jumped, walked,* and *wanted* (Cazden, 1968; Brown, 1973). And Leopold (1949) reported that his daughter, aged 3;3, one day amused herself by adding the diminutive *-ie* ending to all sorts of English words not usually so modified, e.g., wall-*ie*, chair-*ie*, lap-*pie*, and books-*ies* (vol. 4, p. 45).

Rule use, however, requires that one distinguish between implicit knowledge for everyday use and awareness. The addition of a word ending, it could be argued, is simply a matter of everyday usage. Children are always learning new words. However, deciding out of context which endings can be applied presumably does require some level of awareness. Yet only when children begin to make explicit comments on word endings or irregular paradigms can we claim unequivocally that they are reflecting on their language. For instance, Gleitman et al. (1972) reported the following question from a four-year old (p. 139):

Mommy, is it AN A-dult or A NUH-dult?

Leopold (1949) noted a similar question from his daughter, aged 3;11:

You know, Grandpa says 'yourn', 'This is yourn.' Why does he do that?

Hildegard's observation about her grandfather was correct, but, interestingly, she had not seen him for several weeks. A few months later, she pointed out how one distinguished singular and plural forms in German, as follows (vol. 4, p. 61):

If there is one, you have to say *Schuh*; if there are two you have to say *Schuhe*.

Such explicit comments about morphology, however, are not very common.

## CORRECTING ONESELF AND OTHERS

Children make both spontaneous and prompted corrections or repairs to their speech from a very early age. In order to make any repair, children must be able to reflect on their utterances so as to work out what has to be repaired on any one occasion. Repairs seem to spring, for the most part, from a concern with successful communication.

### Spontaneous and Prompted Repairs

Spontaneous repairs appear very early. Both Bohn (1914) and, more recently, Scollon (1976) reported that the very young children they studied repeated words, changing pronunciations until they were able to make themselves understood. The following is a typical example from a girl aged 1;7:

Brenda:  [ǯ]                    (holding up mother's shoe)

          [šI]

          [š]

          [šIš]

          [šu]

          [šu?]

          [šuš]

Mother:  Shoes!    (Scollon, 1976, p. 150)

With each repair, Brenda got closer to the adult form *shoes*; eventually her mother recognized the word she was trying to say.[1]

Slightly older children repair word endings, word order, and even their word choices. Zakharova (1958/1973), for example, found that preschool Russian children often experimented with case endings for unfamiliar words, pronouncing them out loud, and trying out several different endings before deciding on one. And, she noted, "observations of children's speech show that one may encounter, in the younger preschool children, independent corrections of grammatical forms constructed from familiar words as well" (p. 284). Leopold (1949) reported one repair of a word from Hildegard, aged 5;4, speaking German (vol. 4, p. 114):

Zweimal, das is das dreite--das dritte.

Reports of repairs to word order and to word choices appear in a number of studies. Snyder (1914) noted several different types of repair in the speech of a two-and-a-half year old, including addition of a modifier, e.g.,

Dat water--dat dirty water.

change of a word, e.g.,

Might take paddle out boat--might take paddle out canoe.

and change of word order, e.g.,

Down sand beach I been--I been down sand beach.

Many of the spontaneous repairs young children make, then, seem to be motivated by their wish to make themselves understood. They modify their pronunciation, correct their word choice, add modifiers or clarifying phrases, and alter the syntactic form of their utterances. But some repairs may be repairs to that part of the system they are most conscious of at a particular moment. For example, a child working out past tense endings may temporarily be more sensitive to mistakes on those endings.

Several investigators have also suggested that prompted repairs--those *requested* by the listener--may play a critical role in the process of acquiring a language. These repairs force the child to examine what he has just said and identify the source of misapprehension (e.g., Stokes, 1976, 1977; Cherry, in press; Garvey, in press). This is illustrated by the following exchange between three children playing "catch" in a swimming pool (Jefferson, 1972). Steven, aged six is *It* and is counting up to ten while Susan and Nancy, both eight-year olds, are "hiding" by swimming to the middle of the pool:

> Steven: One, two, three [pause] four, five, six [pause] eleven, eight, nine, ten.
>
> Susan : *El*even? --eight, nine, ten?
>
> Steven: Eleven, eight, nine, ten.
>
> Nancy : *El*even?
>
> Steven: Seven, eight, nine, ten.
>
> Susan : That's better.
>
> (Whereupon the game resumes). (p. 295)

The additional stress placed on *eleven* by both eight-year olds presumably helped tell Steven which word he had to correct. Adults and older children who ask for repairs force younger children to reflect on what they are saying and how they are saying it. Requests for repairs, then, may be an important factor in making children monitor their own language.

## Correcting Others

From about four years on, children also seem to become more aware of the "mistakes" of younger brothers and sisters. They comment on them, ask others the reason for them, and often attempt (not always successfully) to correct them. Consider the following comment from Anthony, aged 5;4, about Michael, aged 2;4:

> Michael: Record 'top. Mine!
>
> Anthony: Mike says only *top* instead of *stop*.
>
> (Weir, 1966, p. 164)

On other occasions, Anthony corrected his other brother, David, aged 3;7, for instance by supplying a missing syllable and, in doing so, saying it with extra stress:

> David : I don't have a raser, Antony. I don't have dis.
>
> Anthony: David, you need a *eraser.*
>
> (p. 165)

These explicit comments and corrections can only stem from a growing awareness on Anthony's part of pronunciation and form in English.

*SUPPLYING APPROPRIATE INTERPRETATIONS*

Another situation that may tap children's awareness of language is one in which they are asked to supply interpretations for words and sentences. This they can do by giving definitions or providing paraphrases.

Studies of word definition have usually been designed to trace changes in the complexity of definitions being offered at different ages. Most of them have involved asking children questions like "What is an X?" or "What does X mean?" for terms like *father, south, hole, cup,* etc. (see Piaget, 1926, 1928; Stern & Stern, 1928; Leopold, 1949; Asch & Nerlove, 1960; Wolman & Barker, 1965; Chukovsky, 1968; Al-Issa, 1969; Haviland & Clark, 1974; Andersen, 1975). The earliest definitions offered tend to be phrases in which the relevant words commonly occur: for example, for *hole,* one hears "dig a hole" or "a hole is to dig," and for *cup* one usually hears "what you drink out of" from most children under five or six. As Bolinger (1976) has pointed out, what both children and adults usually do when asked for a definition is first come up with some common phrase in which the word is used. It is only as children get older that they begin to give more complex (and vaguer) dictionary-like definitions. For instance, nine-year-olds may define *cup* as "a curved shaped object for drinking out of, with a hole in the top--to hold in your hand" or as something that will "hold stuff, cup-shaped, sometimes has a handle--sometimes it can be a mug" (Andersen, 1975, p. 97).

But the tendency to offer common phrases or collocations makes definitions a rather indirect route for tapping children's awareness of word meanings. Indeed, giving a definition itself has several prerequisites. The child has to know what a definition is and, on top of that, what constitutes a good definition (see Litowitz, 1977; Johnson-Laird & Quinn, 1976). These complexities make word definitions appear less useful for the study of awareness.

Some investigators have elicited definitions, or rather paraphrases, less directly. Gleitman et al. (1972) asked children aged five and over to make judgments about different sentences--whether they were all right as sentences of English and whether they meant the same thing as other sentences suggested by the experimenter. In making their judgments and explaining them, even the younger children frequently came up with appropriate paraphrases for the sentences in question (see also Hakes et al., in press).

Other studies seem to have elicited paraphrases by accident.  For instance, Slobin and Welsh (1973) found that some types of relative clause given
to a two-year old to imitate were nearly always repeated back in paraphrased
form, e.g.,

Mozart who cried came to my party → Mozart cried and he came to my party;
The owl who eats candy runs fast → Owl eat a candy and he run fast;
The man who I saw yesterday got wet → I saw the man and he got wet.

However, there have been few systematic attempts to get children to give alternative renditions or paraphrases to find out which sentences they consider
to be related in meaning and which not (but see further Grimm, 1975; Smith,
1974).

## ANALYZING LANGUAGE INTO LINGUISTIC UNITS

Most research done on children's ability to isolate particular linguistic
units has been done in connection with reading.  The ability to identify
words, syllables, and individual sounds has long been considered helpful and
even essential in learning how to read (e.g., Francis, 1973).  Children have
been asked to identify words, morphemes, syllables, and sound segments (e.g.,
Bogoyavlenskiy, 1957/1973; Zhurova, 1964; Reid, 1966; Rozin, Poritsky &
Sotsky, 1971; Holden & MacGinitie, 1972; Kingston, Weaver, & Figa, 1972;
Gleitman & Rozin, 1973; Fox & Routh, 1975).

The techniques used to elicit such information often seem to have been
too complicated for children to follow and, as a result, many investigators
have concluded that children do not have the skill to break utterances down
into different units until age six or even later (e.g., Bruce, 1964; Holden
& MacGinitie, 1972).  More recently, however, Fox and Routh (1975) devised
a way of eliciting such information from children as young as three or four
with considerable success.  What they did was ask children to repeat progressively smaller and smaller "bits" of sentences given them by the experimenter.
Given *Peter fell,* for example, the children would be asked to "say just a
little bit of it," namely just *Peter* or just *fell*.  With this technique, Fox
and Routh found that four-year olds were almost perfect at breaking sentences
down into words and syllables.  Segmenting syllables into sounds was much
harder, although there was improvement with age (see also Liberman, 1973).

Some children seem to discover segmentation into morphemes or syllables
quite spontaneously.  One of Grégoire's (1947) sons at the age of four commented that the word *méchanceté* was made up of the stem *méchant* ('naughty')

Etymology →
homophony →

plus an ending, which he pronounced "stê," for which he then demanded a
meaning. He also asked whether there wasn't a word _gentistê from gentil_
('nice'), plus _stê_ -- to parallel _méchanceté_ in both form and meaning (its
opposite). Besides trying to construct similar word forms, children also
use their observations of whole or partial homophony to make up etymologies
for words. Leopold (1949) reported that Hildegard related _sand_ and _sandal_
on this basis (see also Kaper, 1959). Some children spontaneously discover
how to break up syllables into their component sounds and how to sound out
new words (see, for example, Slobin, this volume). Children who learn to
write before they are taught to read also seem to be aware of this level of
analysis and work out very reasonable letter-to-sound correspondences (see
further Read, this volume).

Another domain of investigation that belongs under this heading is that
concerned with what children think words are. This question has been tackled
in several ways. Piaget (1929) originally approached it by asking when children became aware of the arbitrary connection between words and their referents (see also Vygotsky, 1962; Markman, 1976). More recently, Papandropoulou
and Sinclair (1974) have tackled it by asking children questions like "What
is a word?," "How do you know whether something is a word?," "Say a long/
short/difficult word," and "Say a word you have invented [sic] yourself."
By five or so, children answer the first question by describing a word as
"what you use to talk about something" or "what you use to name something."
At the same time, they have difficulty providing instances of long, short,
or difficult words and usually draw only on nouns as examples. By six or
seven, children are often able to identify words explicitly as parts of larger
meaningful expressions (see further Berthoud-Papandropoulou, 1976; Johnson,
1976).

The ability to analyze utterances into smaller and smaller linguistic
units and the ability to identify and talk about those units changes with age.
But, as these recent studies have shown, the actual questions one asks children about such knowledge are critical. Many past studies may have seriously
underestimated how aware children are of different linguistic units and what
they know about both their structure and function in language.

_PRACTICING AND PLAYING WITH LANGUAGE_

Children show another type of awareness in practicing language. They
repeat set sentence frames, substituting one word for another; they try out

different sentence types; they practice newly acquired sounds and words, saying them over and over again, and so on.  At times, such practice may be difficult to distinguish from actual language use, but one possible criterion is whether the child is directing his utterance to an addressee with any possible communicative intent.  In the examples cited below, where the children were practicing sentence frames, there was no addressee:

*Practice*

(1)  Child aged 1;9: Daddy walk on grass / R [child's name] walk on grass--no / Daddy walk on grass--yes / Daddy walk on snow / snow deep / know that word (Bohn, 1914, p. 586).

(2)  Child aged 2;9: Train go on track / car go on track / wheel go on track / little wheel go on track (Snyder, 1914, p. 421).

Practice like this, where children repeat syntactic or phonetic patterns, was examined in great detail by Weir (1962) who recorded her two-and-a-half year old's monologues produced when he was alone in his room before falling asleep.  A typical practice session might involve changing from one sentence type to another (e.g., from negations to questions to statements), substituting pronouns, adjectives, nouns, and verbs in the appropriate slots, producing question-and-answer sequences, and with all this recapitulating some of the day's events.  The following is a fairly typical extract: *Typical practice*

Not a yellow blanket / the white / white / what color / what color blanket / what color mop / what color glass / put on a blanket / white blanket / and yellow blanket / where's yellow blanket / there's a hat / there's another / there's hat / there's another hat / that's a hat / there is the light / where is the light / here is the light (Weir, 1962, p. 112).

Jespersen (1922) observed Danish children producing similar bedtime monologues where they practiced sentences with frequent substitutions of words.  He also noted some daytime practice.  One child, for instance, practiced plural endings as he turned the pages of a book (see also Johnson, 1928).

Children also practice pronunciation, especially of newly acquired sounds.  One evening Weir's child produced the following fragment of a monologue:

*NB*
*Really?*

Back please / berries ([bɛ́rɪz]) / *not* barries ([bǽrɪz]) / barries, barries / *not* barries / berries / ba ba. (p. 108)

This practice occurred just after the child had first seemed to become aware of the contrast between the two vowels, triggered perhaps by his intense

interest in **strawberries**. Another child tried out his newly mastered [r] as
follows:

> Stoly / stoly here / want a stoly / Dave, stoly / story, story / story's
> / de hat / story's de big hat / story's a hat
>
> (Weir, 1966, p. 163).

Trying to get the sounds right like this seems to start very early. Bohn
(1914, p. 579) reported that his child would consciously "repeat a word many
times in an effort to pronounce it correctly" from the age of 1;7 on.
Grégoire (1947) made similar observations, as did Scollon (1976). This form
of practice is clearly relevant to the repairs that even very small children
make in their efforts to be understood. Indeed, it might be hard to distinguish practice from repairs at times, except for the presence of an addressee
and the child's obvious desire to communicate--both of which would suggest
that repeated attempts at a word in the presence of a listener should be
labelled as repairs. Nevertheless, it is clear that practice and repairs
both require some monitoring on the child's part of what he has just said.

## Rhymes, Puns, and Riddles

Practice is often a form of play with language, and exchanges of nonsense
may exploit different properties of the language being acquired. Children
make up rhymes, change the vowels in nonsense syllables, imitate each other
but alter the sounds, intonations, or even stress, and maintain long playful
dialogues (e.g., Johnson, 1928; Chukovsky, 1968; Cazden, 1974; Teece, 1976;
Garvey, 1977). This early play also shows up in rhymes that accompany many
children's games (Opie & Opie, 1959). These game rhymes, surprisingly uniform from place to place, are used in many different play activities--for
counting-out, giving crooked answers, jeering, parodying nursery rhymes and
hymns, and tricking the listener and leading him astray, to list only a few.

Some of the games embodied in such rhymes revolve around puns and riddles. Several observers have collected examples of those popular at different
ages and tried to find out what changes take place in children's appreciation
and understanding of these linguistic games (e.g., Chukovsky, 1968; Herron &
Sutton-Smith, 1971; Goodnow, 1972; Shultz, 1974; Shultz & Horibe, 1974; Aimard,
1975; Sutton-Smith, 1976; Sanches & Kirschenblatt-Gimblett, 1976). These are
their findings.

The commonest riddles, according to Sutton-Smith (1976), are homonymic,
requiring an implicit reclassification of the words in the initial question,

as in:

Q. Why did the dog go out in the sun?

A. He wanted to be a hot dog.

In this example, a class (dogs) and a class attribute suggested by the question (namely, dogs exposed to high temperatures) are reclassified in the answer to form a new class (something edible). This type of riddle makes up about sixty percent of those elicited from children between six and fourteen.

Other riddles require explicit reclassification, as in:

Q. What has an ear but cannot hear?

A. Corn.

In these, a classification is presented and then one of its criterial attributes is negated. Other riddles depend on multiple classifications, as in the following:

Q. What is the difference between a teacher and an engineer?

A. One trains the mind, the other minds the train.

In riddles like these, the homonym is often a double one. Such riddles seem particularly common in French, perhaps because there are so many homonyms available. The riddles or devinettes collected by Migeon and Baslaar-Hoevenaars (1976), for example, are virtually all homonymic, and many involve multiple classifications as well, e.g.:

Q. Quelle différence y-a-t-il entre la Tour Eiffel et une chemise portée pendant huit jours?

A. La Tour Eiffel est colossale et la chemise est sale au col.

Q. Pourquoi le paysan n'est-il pas fumeur de pipe?

A. Parce que la pipe on doit la bourrer avant de fumer et que le paysan doit fumer avant de labourer.

Children's ability to appreciate and make up puns and riddles seems to mark an awareness of language in yet another way: they must be able to recognize potential reclassifications and realize that two sequences with the same sounds can have quite different meanings.

*Figurative Language*

A later skill that may require even more reflection is the production and understanding of figurative speech. To use language metaphorically, children have to be able to extend terms from one domain to another (e.g., apply adjectives for physical dimensions to psychological ones), make appropriate comparisons between qualities or entities in different domains (as in the metaphor, *The man is a stone*), take the part for the whole, and so on. Many of these operations require reflection in deciding whether a metaphorical expression is appropriate or not (e.g., Gardner, Kircher, Winner, & Perkins, 1975), whether a term can be extended to other domains (e.g., Gardner, 1974; Winner, Rosenstiel, & Gardner, 1976; Gentner, 1977), and why some metaphors are appropriate and others not (Gardner et al., 1975).

The ability to reflect on figurative language has been used as one measure of cognitive development (e.g., Billow, 1975; Smith, 1976), and earlier studies generally assumed that children did not understand metaphorical language until the age of ten or so (e.g., Asch & Nerlove, 1960). However, somewhat younger children produce appropriate metaphors on occasion (Gardner et al., 1975; Gentner, 1977) and even give reasons for why a metaphor is appropriate when the figure of speech involves qualities of objects they know well (Winner & Gardner, 1977). Using figurative language, though, seems to demand a slightly different kind of reflection from that exercised in puns and riddles. It requires that children look for analogies and even unexpected parallels among real-world relationships. Puns and riddles, on the other hand, seem to be more firmly grounded in children's everyday exposure to language.

Practicing and playing with language, whether trying out newly acquired sounds or sentence patterns, putting together syllables or words that rhyme, or making up riddles, reflect a growing capacity to use language outside a directly communicative context. Figurative language stands a little to one side in this respect because it is communicative but requires a deliberate stretching of the language to evoke new associations and images. A different kind of sophistication and knowledge seems to be needed to exploit the figurative resources of a language to the full.

## TYPES OF AWARENESS

This preliminary taxonomy has presented some of the sources that might establish the degree to which children are aware of and can actively reflect on the different properties of language. Some of the areas I have reviewed probably have more potential than others for systematic investigations into children's conceptions of language. But which hold the most promise for future study? I have tried to provide a preliminary answer by discussing the types of evidence already available from naturalistic observations, interviews, and experimental studies pertinent to linguistic awareness. In doing this, I have taken "awareness" in its broadest sense and illustrated several different types of awareness. But defining more precisely what awareness is is what I want to turn to next.

The notion of awareness has been studied in other areas of development under the label *metacognition*. Flavell (1976, p. 232) has characterized this as one's knowledge about one's own cognitive processes and their products:

> "I am engaging in metacognition (metamemory, metalearning, meta-attention, metalanguage, or whatever) if I notice that I am having more trouble learning A than B; if it strikes me that I should double-check C before accepting it as a fact; if it occurs to me that I had better scrutinize each and every alternative in any multiple choice situation before deciding which is the best one; if I sense that I had better make a note of D because I may forget it...."

The major concern of these developmental studies has been with the regulatory role of metacognition in remembering, learning, attending, or carrying out actions, in order to achieve particular goals (e.g., Piaget, 1974; Kreutzer, Leonard, & Flavell, 1975; Flavell & Wellman, 1976; Flavell, 1977; Brown, 1978). Brown and DeLoache (in press) have listed a number of metacognitive skills that indicate awareness: predicting the consequences of an action or event, checking the results of one's own actions (did *X* work?), monitoring one's ongoing activity (how am I doing?), testing for reality (does *X* make sense?), and coordinating or controlling deliberate attempts to learn and to solve problems. Do children call on these same skills when it comes to language? In general, the answer seems to be "yes," although there is at least one additional skill that has to be added in classifying types of linguistic awareness, and that is the ability to reflect on the product of an utterance.

The different types of awareness uncovered by the present review are classified in Table 1 under the particular metacognitive skills involved. These skills have been listed roughly in their order of emergence, from most to least basic. The first is the ability to *monitor* one's own ongoing utterances. This activity is a prerequisite for spontaneous repairs, practice, and adjustments of one's speech style to different listeners. Another skill is the ability to *check* the result of one's utterance. Even very young children check to see if the listener has understood, and if not, try again. Rather later, they start to comment explicitly on their own utterances and

Table 1

## METACOGNITIVE SKILLS AND AWARENESS OF LANGUAGE

---

1. *Monitoring one's ongoing utterances*
    (a)  Repairing one's own speech spontaneously
    (b)  Practicing sounds, words, and sentences
    (c)  Adjusting one's speech to the age and status (and language spoken) of the listener

2. *Checking the result of an utterance*
    (a)  Seeing whether the listener has understood or not (and then repairing when necessary)
    (b)  Commenting on the utterances of oneself and others
    (c)  Correcting the utterances of others

3. *Testing for reality*
    (a)  Deciding whether a word or description works or not (and if not, trying another)

4. *Deliberately trying to learn*
    (a)  Practicing new sounds, words, and sentences
    (b)  Role-playing and "doing the voices" for different roles

5. *Predicting the consequences of using inflections, words, phrases or sentences*
    (a)  Applying inflections to "new" words out of context
    (b)  Judging, out of context, which utterance would be politer, or which more appropriate for a specific speaker
    (c)  Correcting word order and wording in sentences earlier judged "silly"

6. *Reflecting on the product of an utterance*
    (a)  Identifying linguistic units (phrases, words, syllables, sounds)
    (b)  Providing definitions
    (c)  Constructing puns and riddles
    (d)  Explaining why certain sentences are possible and how they should be interpreted

---

on those of others. They also correct others. Another skill is *reality testing*: children check on whether a particular word or phrase has "worked" in the sense of getting the listener to understand what they were saying. This testing has a counterpart in prompted corrections where parents at times insist on veridical reference but don't bother to correct grammatical forms (see Clark & Clark, 1977, for a review). A fourth metalinguistic skill is that underlying deliberate attempts to *learn* language. Children practice not only sounds and sentence structures but also the speech styles character-istic of different roles.

The last two skills listed in Table 1 seem to emerge rather later than the others. In *predicting* the consequences of using particular forms, chil-dren use language or make judgments about it out of context. They supply the appropriate inflections to indicate plural, past tense, or diminutive; they judge utterances as appropriate to particular settings or speakers; and they correct sentences that are "wrong." Finally, in *reflecting* on the pro-duct of an utterance, children may be doing something that is never called for in other forms of metacognition. With language, it is possible to re-flect on language structure independent of its actual use. Children iden-tify specific linguistic units--anything from a sound up to a sentence; they provide definitions of words; they construct puns and riddles, and exploit other forms of verbal humour; and they explain why some sentences are possible and how they could or should be interpreted. The classification in Table 1, then, includes all the instances of awareness considered earlier.

With age, children show increasing awareness of language. From the start, they seem to be aware of both form and function. In monitoring what they say, they make spontaneous repairs and practice sounds and sentences from as early as one-and-a-half or two. They also check on whether their listeners have understood them. A little later, by three or four, they begin to adjust their speech style to their listeners, comment on their own utterances and even comment on what they can't do (Smith, 1973; Markman, 1977), ask the occasional question about linguistic forms, and begin to correct other speak-ers. They also practice different speech styles through role-playing, and grow progressively more sensitive to what their listeners will and won't be able to understand (Clark & Clark, 1977). In Table 1, I have tried to cap-ture this growth of awareness by ordering the different phenomena approxi-mately from simpler to more complex under each metacognitive skill. Given the incomplete nature of the observations currently available, however, this ordering is necessarily very tentative.

Earlier, I alluded to the difficulty in distinguishing at times between everyday language use and implicit knowledge where some degree of prediction may be playing a role. Consider the acquisition of the past tense inflection in English. One could argue that in order to apply the past tense ending, two-year olds must be aware of it at some level to identify and select it rather than other possible verb endings to denote completed actions. But not until five or so at the earliest do children appear able to identify the *-ed* ending explicitly as the linguistic unit that adds a past time meaning or make judgments about the appropriate past tense forms of strong verbs like *bring* (see Slobin, this volume). Similarly, they show implicit knowledge of different linguistic units—words, syllables, and phonetic segments —long before they can reflect on those same units explicitly. Implicit knowledge, then, bears some resemblance to Vygotsky's (1962) first stage in the acquisition of knowledge—virtually automatic, unconscious acquisition. This contrasts with the later gradual increase in active, conscious control over knowledge already acquired—Vygotsky's second stage. The latter, roughly speaking, is what is represented in Table 1.

Children's explicit judgments about language presumably depend on the depth of their knowledge at each age. But do all children follow a regular progression in their reflections? Do some children concentrate on one aspect of language where others concentrate on another? One's experience with language presumably has some effect here. Learning two languages at once, for instance, might heighten one's awareness of specific linguistic devices in both. We already know that adult speakers may differ considerably in their ability to reflect on language (e.g., Gleitman & Gleitman, 1970), and these differences presumably correspond to how far children have happened to go in becoming aware of the language they speak.

*CONCLUSION*

The question I have left for last is perhaps the most important: What role does children's awareness play in acquisition itself? Children start out with a very elementary version of language and, to move on to the next stage, must realize that it is inadequate or "wrong." They must therefore become aware of when their language "fails." This means that they have to check the results of their utterances and, when necessary, repair them. Evidence that children do just that spontaneously is already available. Their earliest repairs are of single word utterances that they repeat, with changing pronunciation, until the listener understands what they are trying to say. Later, they repair more than single words: they select more exact descriptions for picking out referents, correct their word order, select the right speech style (and even the right language) for the listener, and so on. And, as early parental requests for clarification show, such repairs can also be elicited. Systematic study of repairs, then, seems likely to prove an important source of information about linguistic awareness.

The other phenomena reviewed here show less promise for discovering linguistic awareness--at least in very young children where awareness may be most critical to acquisition. Judgments of appropriateness, complexity, and form are next to impossible to elicit from very young children and therefore tell us too little too late. The same is true of supplying appropriate interpretations out of context, and of applying rules to new instances. Analyzing language into explicit units also emerges rather late, possibly because children have first to learn a vocabulary for talking about how they use language (see Slobin, this volume). Practice appears much earlier. But because it requires memory for what utterances sound like and the units they consist of, it is probably not as informative as repairs are, even though it emerges at about the same age. All these phenomena, however, are probably critical for filling in our picture of what children are aware of as they get older.

The task from here on, then, is not just to find out what children are aware of, but to go further: first, to find out how awareness itself develops and what relations there are between different types of awareness, and secondly, to establish what role awareness plays in the acquisition of language. The present discussion, I hope, represents a first step in that direction.

*ACKNOWLEDGEMENTS*

The preparation of this paper was supported in part by the National Science Foundation, Grant No. BNS 75-17126. I thank Ellen Markman and Werner Deutsch for their helpful discussion, and Herbert H. Clark for his detailed comments on an earlier version of the manuscript.

*FOOTNOTE*

1. Repetitions and repairs generally are all too often "edited out" of transcripts of children's speech, and they are even harder to catch in paper-and-pencil recordings.

*REFERENCES*

Aimard, P. *Les jeux de mots de l'enfant*. Villeurbanne: Simep-Editions, 1975.

Al-Issa, I. The development of word definitions in children. *Journal of Genetic Psychology*, 1969, 114, 25-28.

Andersen, E.S. Cups and glasses: Learning that boundaries are vague. *Journal of Child Language*, 1975, 2, 79-103.

Andersen, E.S. *Learning to speak with style: A study of the sociolinguistic skills of children*. Unpublished doctoral dissertation, Stanford University, 1977.

Asch, S.E. & Nerlove, H. The development of double function terms in children. In: B. Kaplan & S. Wapner (eds.), *Perspectives in psychological theory*. New York: International Universities Press, 1960.

Bates, E. *Language and context: The acquisition of pragmatics*. New York: Academic Press, 1976.

Berko, J. The child's learning of English morphology. *Word*, 1958, 14, 150-177.

Berko Gleason, J. Code switching in children's language. In: T.E. Moore (ed.), *Cognitive development and the acquisition of language*. New York: Academic Press, 1973.

Berthoud-Papandropoulou, I. *La réflexion métalinguistique chez l'enfant*. Unpublished doctoral dissertation, University of Geneva, 1976.

Billow, R. A cognitive developmental study of metaphor comprehension. *Developmental Psychology*, 1975, 11, 415-423.

Bogoyavlenskiy, D.N. The acquisition of Russian inflections (1957). Translated in C.A. Ferguson & D.I. Slobin (eds.), *Studies of child language development*. New York: Holt, Rinehart & Winston, 1973.

Bohn, W.E. First steps in verbal expression. *Pedagogical Seminary*, 1914, 21, 578-595.

Bolinger, D.L. Meaning and memory. *Forum Linguisticum*, 1976, 1, 1-14.

Brandenburg, G.C. The language of a three-year old child. *Pedagogical Seminary*, 1915, 22, 89-120.

Brown, A.L. Knowing when, where, and how to remember: A problem of metacognition. In: R. Glaser (ed.), *Advances in instructional psychology*. Hillsdale, N.J.: Lawrence Erlbaum Associates, 1978 (in press).

Brown, A.L., & DeLoache, J. Skills, plans, and self-regulation. In: R. Siegler (ed.), *Children's thinking: What develops*. Hillsdale, N.J.: Lawrence Erlbaum Associates, in press.

Brown, R. *A first language: The early stages*. Cambridge, Mass.: Harvard

University Press, 1973.

Bruce, D.J. The analysis of word sounds by young children. *British Journal of Educational Psychology*, 1964, 34, 158-170.

Cazden, C.B. The acquisition of noun and verb inflections. *Child Development*, 1968, 39, 433-448.

Cazden, C.B. Play and metalinguistic awareness: One dimension of language experience. *The Urban Review*, 1974, 7, 28-39.

Cherry, L. The role of adults' requests for clarification in the language development of children. In: R.O. Freedle (ed.), *Discourse processing: Multidisciplinary perspectives*, vol. 2. Hillsdale, N.J.: Ablex Publishing Company, in press.

Chukovsky, K. *From two to five*. Berkeley & Los Angeles: University of California Press, 1968.

Clark, H.H., & Clark, E.V. *Psychology and language*. New York: Harcourt Brace Jovanovich, 1977.

Edelsky, C. Acquisition of an aspect of communicative competence: Learning what it means to talk like a lady. In: S. Ervin-Tripp & C. Mitchell-Kernan (eds.), *Child discourse*. New York: Academic Press, 1977.

Flavell, J.H. Metacognitive aspects of problem solving. In: L.B. Resnick (ed.), *The nature of intelligence*. Hillsdale, N.J.: Lawrence Erlbaum Associates, 1976.

Flavell, J.H. Metacognitive development. Paper presented at the NATO Advanced Study Institute on Structural/Process Theories of Complex Human Behavior, Banff, Alberta, Canada, June 1977.

Flavell, J.H., & Wellman, H.M. Metamemory. In: R.V. Kail, Jr., & J.W. Hagen (eds.), *Perspectives on the development of memory and cognition*. Hillsdale, N.J.: Lawrence Erlbaum Associates, 1977.

Fox, B., & Routh, D.K. Analyzing spoken language into words, syllables, and phonemes: A developmental study. *Journal of Psycholinguistic Research*, 1975, 4, 331-342.

Francis, H. Children's experience of reading and notions of units in language. *British Journal of Educational Psychology*, 1973, 43, 17-23.

Gardner, H. Metaphors and modalities: How children project polar adjectives onto diverse domains. *Child Development*, 1974, 45, 84-91.

Gardner, H., Kircher, M., Winner, E., & Perkins, P. Children's metaphoric production and preferences. *Journal of Child Language*, 1975, 2, 125-141.

Garvey, C. Play with language and speech. In: S. Ervin-Tripp & C. Mitchell-Kernan (eds.), *Child discourse*. New York: Academic Press, 1977.

Garvey, C. Contingent queries and their relations in discourse. In: E.O. Keenan (ed.), *Studies in linguistic pragmatics*. New York: Academic Press, in press.

Gelman, R., & Shatz, M. Appropriate speech adjustments: The operation of conversational constraints on talk to two-year olds. In: M. Lewis & L. Rosenblum (eds.), *Interaction, conversation, and the development of language*. New York: Wiley, 1977.

Gentner, D. Children's performance on simple spatial metaphors. Paper presented at the Ninth Annual Child Language Research Forum, Stanford University, March 1977.

Gleitman, L.R., & Gleitman, H. *Phrase and paraphrase: Some innovative uses of language*. New York: Norton, 1970.

Gleitman, L.R., Gleitman, H., & Shipley, E.F. The emergence of the child as grammarian. *Cognition*, 1972, 1, 137-164.

Gleitman, L.R., & Rozin, P. Teaching reading by use of a syllabary. *Reading Research Quarterly*, 1973, 8, 447-483.

Goodnow, J. Rules and repertoires, rituals and tricks of the trade: Social and informational aspects to cognitive and representational development. In: S. Farnham-Diggory (ed.), *Information processing in children*. New York: Academic Press, 1972.

Grégoire, A. *L'apprentissage du langage* (2 vols.). Paris: Droz, 1947.

Grimm, H. Analysis of short-term dialogues in 5-7 year olds: Encoding of intentions and modifications of speech acts as a function of negative feedback. Paper presented at the Third International Child Language Symposium, London, September 1975.

Hakes, D.T., Evens, J.S., & Tunmer, W. The emergence of linguistic intuitions in children. *Monographs of the Society for Research in Child Development*, in press.

Haviland, S.E., & Clark, E.V. 'This man's father is my father's son:' A study of the acquisition of English kin terms. *Journal of Child Language*, 1974, 1, 23-47.

Herron, R.E., & Sutton-Smith, S. (eds.), *Child's play*. New York: Wiley, 1971.

Holden, M.H., & MacGinitie, W.H. Children's conceptions of word boundaries in speech and print. *Journal of Educational Psychology*, 1972, 63, 551-557.

James, S.L., & Miller, J.F. Children's awareness of semantic constraints in sentences. *Child Development*, 1973, 44, 69-76.

Jefferson, G. Side sequences. In: D.N. Sudnow (ed.), *Studies in social interaction*. New York: Free Press, 1972.

Jespersen, O. *Language: Its nature, development, and origin*. London: Allen & Unwin, 1922.

Johnson, C.N. *Children's reflections on the nature of names*. Unpublished doctoral dissertation, University of Minnesota, 1976.

Johnson, H.M. *Children in the 'Nursery School.'* New York: John Day & Co., 1928.

Johnson-Laird, P.N., & Quinn, J.G. To define true meaning. *Nature*, 1976, 264 (16 December), 635-636.

Kaper, W. *Kindersprachforschung mit Hilfe des Kindes: Einige Erscheinungen der kindlichen Sprachwerbung erläutert im Lichte des vom Kinde gezeigte Interesses für Sprachliches*. Groningen: Wolters, 1959.

Kingston, A.J., Weaver, W.W., & Figa, L.E. Experiments in children's perceptions of words and word boundaries. In: F.P. Green (ed.), *Investi-*

*gations relating to mature reading.* Milwaukee, Wis.: National Reading Conference, Inc., 1972.

Kreutzer, M.A., Leonard, C., & Flavell, J.H. An interview study of children's knowledge about memory. *Monographs of the Society for Research in Child Development*, 1975, 40 (Serial No. 159).

Leopold, W.F. *Speech development of a bilingual child* (4 vols.). Evanston, Ill.: Northwestern University Press, 1949.

Liberman, I.Y. Segmentation of the spoken word and reading acquisition. Paper presented at the Symposium on Language and Perceptual Development in the Acquisition of Reading, held at the Biennial Meeting of the Society for Research in Child Development, Philadelphia, March 1973.

Limber, J. The genesis of complex sentences. In: T.E. Moore (ed.), *Cognitive development and the acquisition of language*. New York: Academic Press, 1973.

Litowitz, B. Learning how to make definitions. *Journal of Child Language*, 1977, 4, 289-304.

Maccoby, E.E., & Bee, H.L. Some speculations concerning the lag between perceiving and performing. *Child Development*, 1965, 36, 367-377.

Markman, E.M. Children's difficulty with word referent differentiation. *Child Development*, 1976, 47, 742-749.

Markman, E.M. Realizing that you don't understand: A preliminary investigation. *Child Development*, 1977, 48, 986-992.

Migeon, D.A., & Baslaar-Hoevenaars, L. Où, pourquoi, comment des devinettes? Etude linguistique de Maîtrise inédite, Université de Provence (Aix-Marseille 1), 1976.

Moore, T.E. Linguistic intuitions of twelve-year olds. *Language and Speech*, 1975, 18, 213-216.

Opie, I., & Opie, P. *The lore and language of schoolchildren*. Oxford: Oxford University Press, 1959.

Papandropoulou, I., & Sinclair, H. What is a word? Experimental study of children's ideas on grammar. *Human Development*, 1974, 17, 241-258.

Piaget, J. *The language and thought of the child*. London: Routledge & Kegan Paul, 1926.

Piaget, J. *Judgment and reasoning in the child*. London: Routledge & Kegan Paul, 1928.

Piaget, J. *The child's conception of the world*. London: Routledge & Kegan Paul, 1929.

Piaget, J. *La prise de conscience*. Paris: Presses Universitaires de France, 1974.

Reid, J. Learning to think about reading. *Educational Research*, 1966, 9, 56-62.

Rozin, P., Poritsky, S., & Sotsky, R. American children with reading problems can easily learn to read English represented by Chinese characters. *Science*, 1971, 171, 1264-1267.

Sachs, J., & Devin, J. Young children's use of age-appropriate speech styles in social interaction and role-playing. *Journal of Child Language*,

1976, 3, 81-98.

Sanches, M., & Kirschenblatt-Gimblett, B.  Children's traditional speech play and child language.  In: B. Kirschenblatt-Gimblett (ed.), *Speech play*.  Philadelphia, Pa.: University of Pennsylvania Press, 1976.

Scollon, R.  *Conversation with a one year old: A case study of the developmental foundation of syntax*.  Honolulu: University Press of Hawaii, 1976.

Scholl, D.M., & Ryan, E.B.  Child judgments of sentences varying in grammatical complexity.  *Journal of Experimental Child Psychology*, 1975, 20, 274-285.

Shatz, M., & Gelman, R.  The development of communication skills: Modifications in the speech of young children as a function of listeners.  *Monographs of the Society for Research in Child Development*, 1973, 38 (Serial No. 152).

Shipley, E.F., Smith, C.S., & Gleitman, L.R.  A study in the acquisition of language: Free responses to commands.  *Language*, 1969, 45, 322-342.

Shultz, T.R.  Development of the appreciation of riddles.  *Child Development*, 1974, 45, 100-105.

Shultz, T.R., & Horibe, F.  Development of the appreciation of verbal jokes.  *Developmental Psychology*, 1974, 10, 13-20.

Slobin, D.I., & Welsh, C.A.  Elicited imitation as a research tool in developmental psycholinguistics.  In: C.A. Ferguson & D.I. Slobin (eds.), *Studies of child language development*.  New York: Holt, Rinehart, & Winston, 1973.

Smith, C.S.  Paraphrase: An exploratory experiment with children 5-7 years old.  *Le Langage et l'Homme*, 1974, 26, 11-19.

Smith, J.W.  Children's comprehension of metaphor: A Piagetian interpretation.  *Language and Speech*, 1976, 19, 236-243.

Smith, N.V.  *The acquisition of phonology: A case study*.  Cambridge: Cambridge University Press, 1973.

Snow, C.E., & Ferguson, C.A. (eds.), *Talking to children: Language input and acquisition*.  Cambridge: Cambridge University Press, 1977.

Snyder, A.D.  Notes on the talk of a two-and-a-half year old boy.  *Pedagogical Seminary*, 1914, 21, 412-424.

Stern, C., & Stern, W.  *Die Kindersprache: Eine psychologische und sprachtheoretische Untersuchung* (4th. edition).  Leipzig: Barth, 1928.

Stokes, W.  Children's replies to requests for clarification: An opportunity for hypothesis testing.  Paper presented at the First Annual Boston University Conference on Language Development, October, 1976.

Stokes, W.  Motivation and language development: The struggle towards communication.  Paper presented at the Biennial Meeting of the Society for Research in Child Development, New Orleans, March 1977.

Sutton-Smith, S.  A developmental structural account of riddles.  In: B. Kirschenblatt-Gimblett (ed.), *Speech play*.  Philadelphia, Pa.: University of Pennsylvania Press, 1976.

Teece, C.  Language and play: A study of the relationship between functions and structures in the language of five-year old children.  *Language and*

*Speech*, 1976, 19, 179-192.

de Villiers, J.G., & de Villiers, P.A. Competence and performance in child language: Are children really competent to judge? *Journal of Child Language*, 1974, 1, 11-22.

Vygotsky, L.S. *Thought and language*. Cambridge, Mass.: M.I.T. Press, 1962.

Weir, R.H. *Language in the crib*. The Hague: Mouton, 1962.

Weir, R.H. Some questions on the child's learning of phonology. In: F. Smith & G.A. Miller (eds.), *The genesis of language*. Cambridge, Mass.: M.I.T. Press, 1966.

Winner, E., & Gardner, H. What does it take to understand a metaphor? Paper presented at the Biennial Meeting of the Society for Research in Child Development, New Orleans, March 1977.

Winner, E., Rosenstiel, A.K., & Gardner, H. The development of metaphoric understanding. *Developmental Psychology*, 1976, 12, 289-297.

Wolman, N., & Barker, E.N. A developmental study of word definitions. *Journal of Genetic Psychology*, 1965, 107, 159-166.

Zakharova, A.V. Acquisition of forms of grammatical case by preschool children (1958). Translated in C.A. Ferguson & D.I. Slobin (eds.), *Studies of child language development*. New York: Holt, Rinehart & Winston, 1973.

Zhurova, L.Ye. The developmental analysis of words into their sounds by preschool children. *Soviet Psychology and Psychiatry*, 1964, 2, 11-17.

# A Case Study of Early Language Awareness

Dan I. Slobin

University of California, Berkeley, CA 94720, USA

Along with the development of language itself, there emerges a capacity to
attend to language and speech as objects of reflection.  The development of
language awareness is, of course, part of the general development of consciousness and self-consciousness. One can distinguish levels of metalinguistic capacity, from the dimly conscious or preconscious speech monitoring
which underlies self-correction, to the concentrated, analytic work of the
linguist.  Much of this route is traversed in the preschool years.  The following aspects of language awareness appear, between the ages of two and six:

(1)  self-corrections and re-phrasings in the course of ongoing speech;
(2)  comments on the speech of others (pronunciation, dialect, language,
       meaning, appropriateness, style, volume, etc.);
(3)  explicit questions about speech and language;
(4)  comments on own speech and language;
(5)  response to direct questions about language.

This paper is a discussion of the development of language awareness in
my daughter, Heida, between the ages of 2;9 and 5;7.  Examples are drawn from
my diary observations of her linguistic development, reflecting the range of
metalinguistic phenomena observable in one preschool child.  Heida lived
abroad between the ages of 2;9 and 3;11--chiefly in Turkey, but with travel
through a number of other countries.  The resulting contact with a series of
foreign languages makes this case different both from normal monolingual and
bilingual development, and may have stimulated particularly early attention
to linguistic phenomena.  I discuss several aspects of this attention below.

## METALINGUISTIC VOCABULARY

It would be valuable to study the language available to children for the discussion of language and speech. Heida used the verb *mean* at an early age, due to contact with foreign languages; however, it was also used to request definitions of English words. By 3;4 the following metalinguistic vocabulary items were attested: *mean, be called, name, word, say, speak, voice,* and *look like* (meaning *sound like*).

### Mean

The use of *mean* was present from the second day in Europe, at age 2;9. At first it was used to elicit pairings of English and foreign words, in either direction. After having been in Czechoslovakia, Germany, Austria, and Yugoslavia, she would initiate series of questions, such as: "What does *bread* mean in German? What does *bread* mean in Yugoslavian?" and so forth. This question frame was later replaced by frames using *say, call,* and *word*: "How say X?", "What do you call X?", "What is X called?", and "What the word for X?". Apparently she understood that naming a language (*German, Turkish*) in one of these question frames would elicit a strange-sounding word which could be used with practical effect in a communicative setting (ordering food, buying things, etc.). (Dictionaries, being the frequent source of such verbal counters, became prized possessions).

She did not understand, however, that her own speech could be part of such a language game. English words could not be distinguished from the concepts to which they make reference. This is most clearly revealed in an observation from age 3;4, after she had been in Turkey for over four months:

§1 (3;4). She doesn't accept her English words as a language, but apparently treats them as something like pure word meaning. She asks: "How a say *red* in English?" She doesn't accept *red* as an answer, but insists on something else to be called an English word, along with words in other languages. Later in the day she asks: "What is spoon called in English?".

In response to questions of the form, "What does X mean?", I would sometimes provide an English definition rather than a foreign word. Heida readily accepted both translations and definitions as responses, and came to use *mean* for both functions in her own speech:

§2 (3;3). She offered the following spontaneous translations, which is correct:"*Gel* means *come* and *koş* means *run*."

§3 (3;3). She has been confused for some time about the meaning of *before* and *after*, and today asked explicitly: "What does *before* mean?"

Note that the same question is used both to request definitions and foreign equivalents.

At the same time she was able to discuss the meanings of her own statements, accepting and rejecting paraphrases as "meaning" what she was trying to say. Somewhat later, when almost 3;6, she was able to offer her own paraphrases:

§4 (3;5). She explains her own idiosyncratic usage, "Even I want some milk." She accepts a paraphrase with *really*, and says it means: *"I want some milk--very much I want some milk."*

It is evident that part of Heida's understanding of *mean* related to the appropriate usage of words, either English or foreign. The requests for definitions and discussions of paraphrases indicate that she was not applying *mean* solely to elicit pairs of English and foreign words. It was not always clear, however, what kind of answer she expected to a "What does X mean" question. Beginning at 3;1 she began to take English words apart, expecting each part to have a "meaning," as if unwilling to accept the existence of duality of patterning:

§5 (3;1). Heida asks:"*Cookie*. What does *cook* mean?" When given an answer, she went on to ask: "What does *ku* mean?" She did this with several other words, e.g., *"Tiger*. What does *ti* mean?" She even dissected ð into *t* and ʒ: "Wit—what does wit mean—wit—witch?" She also broke down *w*: "What does *oo* mean? Wall—*oo...all*."

§6 (3;1). She attempts to break an utterance down into phrases, words, and parts of words: "Are these little petal things? What's *are these*? What's *are*? What's *are* mean? What's *me* mean?"

It was not until 4;7 that she offered a spontaneous definition: "Today I learned what *super* means. It means *really, really, really* something."

## Say

*Say* first appeared in reference to writing, at 3;1. Looking at signs, she would ask, "What does it says?" *Say* was also used as another way of asking for translations: "How say X?" These uses seem tied to the immediate speech situation, but the following example may indicate a generic use of *say* as habitual speech behavior. At 3;1 she was learning to count in Turkish, and seemed to be struck by the arbitrariness of ordering of number words: "Why cause you don't say first *beş, iki, dört (five, two, four)*?"

At 3;4 *say* was used to refer to inner speech: "I said to myself, 'I want my mama and my papa to play with me'."

## Use of Other Metalinguistic Terms

*Speak* was used only in the context of specific language names:

§7 (3;3). She asks: "Do we speak English because we're from Engly?" I explain, and then ask: "Where's German from?" She answers: "Germny... Germy." I go on: "Russian?" "Rushy." "Turkish?" "Turkey." "Italian?" "Talmy."

From the first month in Europe, at 2;10, she noted that some foreign words "sound funny." On one occasion, at 3;3, she created terms to characterize foreign accent (also indicating memory for accent):

§8 (3;3). I was telling a joke in a Yiddish accent and Heida said: "That looks like Great-Grandma" (who speaks with a Yiddish accent). There was no one on the street who looked like her great-grandmother. Heida added: "That's like Great-Grandma's voice." Eight days earlier she had heard a five minute tape recording of her great-grandmother's voice. Apparently this was sufficient for her to recognize the foreign accent.

## Metalinguistic Comprehension

In the context of various informal tests, Heida showed ability to comprehend instructions to attend to features of language or aspects of language use. At 3;5 she was able to play a game with the instruction, "Give me a word that sounds like X." At 4;3 she was able to answer questions about "Which is right to say" in reference to past tense forms (discussed below). At 4;4, with limited reading ability, she could play a category game using first letters as cues, as in, "Give me a food starting with A." At 4;5 she understood *backwards* as an instruction both to spell and to pronounce words backwards. At the same age she was easily trained to understand *opposite* as an instruction to provide antonym responses.

## SPONTANEOUS ATTENTION TO ADULT SPEECH

The examples discussed above of attention to foreign speech and accent indicate that Heida was actively monitoring adult speech. Furthermore, there were many examples of explicit discussion of things which puzzled her, including metaphor, anomaly, synonymy, and unfamiliar words. At 3;2 she was disturbed by apparent synonymy, asking: "Why cause you have two names, *orange* and *tangerine*?", apparently thinking that the two names apply to the same fruit. An observation from 4;2 shows attention to new words:

§9 (4;2). She monitors adult speech closely for unfamiliar words and asks for their meanings—both in speech addressed to her and in over-heard conversations. For example, I say: "She's really tired. Maybe she'll sleep really soundly, and then she won't have any dreams." She asks: "What does *soundly* mean?"

At 4;5 she was struck by an apparent anomaly:

§10 (4;4). While drawing, she overhears an adult conversation in which someone says, "Klee says . . . ." Heida interjects: "Clay doesn't have mouths!"

And at 4;9:

§11 (4;9). She picks up on usage which violates her sense of grammaticality. On the TV news she hears the word *persons* and mulls over it for some time, since she had recently discovered that *people* is the normal plural of *person*.

## SPONTANEOUS ATTENTION TO OWN SPEECH

The lowest level of attention to own speech comes from spontaneous corrections and re-phrasings. An observation from 3;1 suggests that self-monitoring was relatively late to develop:

§12 (3;1). If her verbal formulations are not at once understood, she lies on the floor and cries or screams--but doesn't attempt to reformulate her statement.

This suggestion is supporting by a diary note from 3;2:

§13 (3;2). Self-corrections are still rare, but note: "It's watching we cutting . . . our cutting . . . we cutting . . . It's watching our's cutting."

This level of attention is well established by 3;4; for example:

§14 (3;4). Successive reformulations: "Some friend of mine gave it to me. A girl friend gave it to me. A girl of my friend gave it to me. A girl my friend gave it to me." Self-correction: "You didn't give me a fork. You didn't gave me a fork."

Attention to the sound qualities of words seemed to appear earlier than attention to meaning or grammar. I have already noted spontaneous analysis of words into syllables and sounds, beginning at 3;1. At the same age she engaged in rhyming play, noticing sound similarities in words in her own speech:

§15 (3;1). "Eggs are beggs. Enough—duff. More—bore." Other attention to word details: "It's just the same—tuna tune." She made up the name *hokadin* and broke it into syllables: *hoke—a—din*.

Similar attention to word and sound segmentation appeared about a year later, in connection with acquisition of reading.

At 4;3 she was aware of her own speech articulation skills, noting progress:

§16 (4;3). She clearly repeated "Look at *that*," trying to draw my attention to something, but really trying to draw attention to her first clear pronunciation of *th*.

*knows she's correct! + shows off her th pronunciation.*

*DEGREE OF PERSONAL CONTROL OVER LANGUAGE*

The immediate impact of the foreign language experience on Heida was the introduction of non-English vocabulary. From the second day in Europe, at age 2;9, she invented a new word for milk, insisting that it be called [bap]. She frequently babbled in foreign sounds, and continued to invent words of her own. She clearly had no difficulty in accepting alternate sound patterns as names of things in different languages, and could play the game of asking for translation equivalents from 2;9 on. At the same time, as indicated in the discussion of *mean* above (§1), English words seemed to be exempt from this flexibility of usage. *Bread* really is "bread," though it can be called *Brot* or *hleb* or *ekmek* in certain special languages games. In similar fashion, Heida was troubled by synonymy (orange and tangerine example, above) and rejected metaphor, insisting on literal meanings.

Yet, at the same age (3;4-3;5) she began to take a pretend attitude toward name changes, tentatively willing to unhook word and referent, at least in play:

§17 (3;4). She plays with the idea of changing names with her best friend, Jess: "I wanna be called *Jess*. Sometimes Jess can be called *Jess* and I can be called *Jess*. I will have two names: Heida and Jess."

§18 (3;5). She is wearing pants but wants to be wearing a dress so that she can dance. She says to me: "Call it a dress, please." I reply, "It's not a dress." She says, "Pretend."

It is impossible to know to what extent these uncertain attitudes about the fixed or variable nature of word-referent relationships were due to her multilingual exposure. She actively reflected on this problem, showing a concern with justification for word usage, both within and between languages. Her questions suggest a nascent awareness of the separability of sound vehicle and concept. At 3;2 she questioned both the use of a proper name ("Why cause he name was George?") and a compound noun ("Why cause it's called *Thanksgiving*?"); and at 3;3 she questioned Turkish usage ("Why in Turkish *kaka* is BM?"). Although the data are scanty, these and other observations (especially those on word segmentation) at least suggest that a child of this age is able to reflect on the sound-meaning relationship.

*ORGANIZATION BELOW THE LEVEL OF AWARENESS*

All regularities of speech, of course, reflect underlying structures. Heida's oriented strategies for acquiring foreign vocabulary, however, reflect semantic structures on the level of the lexicon rather than the individual sentence. Two examples are suggestive, one very early, and one much later:

§19 (2;10). [In Prague]  She became fascinated with a little dictionary, which she seemed to see as the key to foreign language.  She easily got the idea of asking me to give Czech words from the dictionary in response to English words, and spent a half hour of concentration asking for words which I gave her.  (After playing the game with Czech words, she then wanted German words, indicating some awareness of different languages.)  She first asked only for nouns, giving me one English object name after another.  Then I suggested "another kind of word, like *walk* or *eat*."  She easily switched to verbs, but soon fell into verb phrases (e.g., "eat some meat").  She could not pick up on adjectives, though.

§20 (3;6). [In Istanbul]  Great increase of interest in Turkish.  Asks how to say things.  Very systematic—e.g., wanted to know how to say: "It's mine, it's yours, it's not mine, it's not yours."  She asked for a variety of locative expressions, placing her finger in a cup, in a hole, on the table, under a plate, etc.  Then asked for pairs of affirmative and negative sentences, e.g., "You spilled your water—You didn't spill your water."

## *AN EMERGING SENSE OF GRAMMATICALITY*

A sense of grammaticality is implicit in self-corrections, and perhaps in puzzlements over unassimilable aspects of adult speech.  In the case of Heida's developing awareness of the English past tense, however, one can trace a path from initial unawareness, to a sense of correctness accompanied by uncertainty in regard to particular words, to an explicit normative sense.

The story begins at 4;2, when Heida's speech was rich in overgeneralizations.  (Presumably these forms had been used unselfconsciously for some time.)  An observation at 4;2 notes: "She adamantly refuses to accept irregular past tense—i.e., the correct forms—insisting on her own, long-term overregularizations."  A sense of appropriateness is already present—but for *her* forms, rather than the adult forms.

A few weeks later, at 4;3, she judges adult forms as correct in a test situation, though she does not use all of these forms herself:

§21 (4;3).  I ask Heida a series of questions of the following form: "Suppose you were eating something yesterday.  Which is right to say: 'Yesterday I ate something' or 'Yesterday I eated something'?"  The order of correct and incorrect verb forms varied.  The sentences frames varied, but all avoided mention of the verb in the past tense.  She accepted the task at once, and almost always gave me one word answers—confidently supplying the correct form of the verb (with two exceptions).  That is, her response to the example was *ate*.  This is unusual in that she rarely uses some of these verbs correctly in her own speech; yet she is clearly aware of the correct forms.  Note that about a month ago she was adamant in defending her overregularizations (e.g., *camed*).  Also, in dictating a letter today, she corrected herself twice: ". . . comed . . . came . . . comed . . . came."

Informal testing of this sort continued, with no feedback, but·probably
drew her attention to discrepancies between her forms and adult forms.  At
4;4 she would change judgments in conflicts:

§22 (4;4).  In the past tense test she accepts correct alternatives and
rejects incorrect alternatives, even if she doesn't use the correct
forms in her speech.  For example, she offers *knowed* but accepts *knew*;
she offers *winned*, rejects *wan*, and accepts *won*.  She will occasionally
change her initial form when challenged ("Are you sure?"  "Is there another way?").  Thus she has a sense of grammaticality which is not regu-
larly reflected in her use of past tense forms.

By 4;5 she began to judge both her forms and the standard forms as correct:

§23 (4;5).  She now accepts several past tense forms as correct for
irregular verbs.  For example, she says *stringed*; I offer *strung*; she
concludes: "*Strung* is OK too."  Apparently she has decided, for now,
that some verbs have equally correct alternate past tense forms.

§24 (4;5).  Her sense of past tenses is becoming more open.  For example,
on the past tense today, she accepts both *finded* and *found*.  I ask: "How
come there are two ways, like *finded* and *found*?  Are they both right?"
She replies: "I don't know.  I *think* they are."  For many verbs, now,
she accepts both forms on the test.

This is a curious intermediate stage in forming an explicit sense of
grammaticality.  It continued for several months, at least until 4;9.  During
this period of concerted attention to the past tense--both spontaneously and
in periodic testing--apparently both forms sounded correct to her.  It is as
if she had a good statistical sense that both standard and overgeneralized
forms occurred frequently, but had failed to note that the overgeneralizations
came from her own speech and the standard forms from the speech of adults.
Perhaps a sense of familiarity with both forms led her to judge both as grammatical (that is, as "right"), suggesting that judgments of acceptability
may be based as much on familiarity as on consistency with norms.  A charming
and rather amazing example from 4;7 graphically reveals the flickering nature
of the sense of grammaticality at this stage.  Overgeneralizations planted
in adult speech elicited protest from Heida only if the standard form happened to be momentarily present in her consciousness:

§25 (4;7).  If she has just used the correct past tense of an irregular
verb, she is annoyed with me if I respond to her with the overregularization; but if she has used the overregularization, she does not object
to my following suit.  If I follow her incorrect form with the correct
form, she will often switch to the correct form.  The following dialogue
is a good example of how the two forms flit in and out of consciousness
in the course of natural conversation:

> Dan  :  Hey, what happened last night after we left?  Did Barbara
>          [the baby sitter] read you that whole story?  Remember you
>          were reading *Babar*?

Heida: Yeah . . . and, um, he . . . she also . . . you know . . .
mama, mama, uh, this morning after breakfast, read[1] the
whole, um, book of the three little pigs and that, you know
that book, that . . .
　　　　　　　　　[digression of about one minute]
Heida: I don't know when she readed . . .
Dan : You don't know when she what?
Heida: . . . she readed the book. But you know that book—that
green book—that has the gold goose, and the three little
pigs, and the three little bears, and that story about the
king?
Dan : M-hm.
Heida: That's the book she read. She read the whole, the whole
book.
Dan : That's the book she *readed* huh?
Heida: Yeah . . . *read*! [annoyed].
Dan : Oh.
Heida: Dum-dum!
　　　　　　　　　[brief interlude about dressing]
Dan : Barbara readed you *Babar*?
Heida: *Babar*, yeah. You know, cause *you* readed some of it too.
Dan : Well I just started it.
Heida: Yeah. She readed all the rest.
Dan : She read the whole thing to you, huh?
Heida: Yeah . . . nu-uh—*you* read some.
Dan : Oh, that's right; yeah, I readed the beginning of it.
Heida: *Readed*?! [annoyed surprise] *Read*! [insisting on the
obvious].
Dan : Oh yeah--read.
Heida: Will you stop that, papa?
Dan : Sure.

Beyond 4;9, she began to accept a single standard of correctness, recognizing
her own overgeneralizations as errors. Perhaps these examples represent a
general phenomenon of the emergence of linguistic norms in various domains.

*LANGUAGE AWARENESS AND READING*

　　　Learning to read requires awareness to several levels of language.
Early attempts to segment words into syllables and small units of sound preceded the acquisition of reading and writing. Detailed phonetic analysis
was reflected in early spelling. I will not explore these issues here, as
they are similar to the phenomena reported in detail by Charles Read (1971).
A few examples indicate this sort of metalinguistic attention:

　§26 (4;4). She tries to spell *pee* and insists that it should be spelled
PHEE, emphasizing the aspiration on P, but also the end-glide on the
vowel, which she shades off into H. She spells *pad* as PD, not feeling
a need for a vowel, but following the tongue as it comes to rest. She
begins to sound out *ice cream* as /a/, and tries to spell it with initial
A; begins to sound out *angel* as /e/.

　§27 (4;4). She can play a category game using first letters as cues, e.g.,

"Give me a food starting with A." In playing this game, she offered *chair* as a response to, "Give me the name of a piece of furniture starting with T."

*CONCLUSION*

These observations are only suggestive of the nature and extent of early language awareness. The capacity to reflect on the form, meanings, and uses of language is clearly present at a very early age. More detailed investigation is needed to establish the generality and sequencing of the metalinguistic abilities reflected in this case study.

*ACKNOWLEDGEMENTS*

This research was supported, in part, by grants from NIMH to the Language-Behavior Research Laboratory and from the Grant Foundation to the Institute of Human Learning, University of California at Berkeley. Special thanks to Heida Slobin, for originally providing the data at an early age, and agreeing to their publication at age ten.

*FOOTNOTE*

1. "Read" represents /red/ throughout.

*REFERENCES*

Read, C. Pre-school children's knowledge of English phonology. *Harvard Educational Review*, 1971, 41, 1-34.

# An Experimental Study of Children's Ideas About Language

Ioanna Berthoud-Papandropoulou

University of Geneva, F.P.S.E., CH-1200 Geneva, Switzerland

From diary studies (see especially Kaper, 1959) as well as recent experiments, it has become clear that children reflect on language well before they receive any formal teaching in grammar: both spontaneously and in response to questions, they make remarks on pronounciation, on morphology, they correct other speakers, they remark on meaning and form, and they may even make puns. But do children also reflect on language in a more "philosophical" way? Do they have ideas about the special properties that make natural language unique as a means of communication and representation? Such metalinguistic activity can be considered within the general framework of cognitive activity for two reasons. First of all, language by its very nature is a product of human cognition as well as a representational system which the child has to reconstruct and learn to use. And secondly, the fact that language becomes an object of human thinking is one manifestation of the general structuring of knowledge that takes place during cognitive development.[1]

Metalinguistic awareness of the kind concerned might be investigated using experimental methods, especially dialogue techniques such as those developed by Piaget. However, the problems that can arise in all psycholinguistic research when aims and means coincide may be especially acute in this kind of research: both the experimenter and the child must *use* language to express their ideas *about* language (Papandropoulou & Sinclair, 1974; Berthoud-Papandropoulou, 1976).

To narrow the scope of such an investigation, a choice needs to be made among basic features of language that might be studied. We decided to concentrate upon two of these: first, the fact that words have no intrinsic relation to the things they stand for, but are just conventional and arbitrary symbols; and second, that words with different meanings are themselves made up of abstract, meaningless phonemes that occur in different orders. These correspond then to Saussure's *arbitraire du signe* on the one hand, and the design feature Hockett (1968) called *duality of patterning* on the other. To find out how children's ideas develop about these aspects of natural language, several experiments were designed, all dealing with a particular linguistic unit, the *word*. This unit was chosen--though it has never been sufficiently defined--because it seems to be more understandable to naive speakers than other, more technical terms such as *syllable* or *sentence*. Words as units have the added advantage that they can be analyzed phonologically, syntactically and semantically; they are made up of smaller units and form part of larger units, and their reference is generally extra-linguistic. Moreover, *word* is probably the most frequently used term in non-technical discussions about language.

Five experiments were carried out to explore various aspects of awareness of words:

-- One study's aim was to obtain children's definitions of the term *word*. In order to probe the children's ideas, we started a conversation that led to the questions: "Do you know what a word is? What do you think?"

-- In another experiment, the experimenter uttered seven or eight simple, frequently used words one by one (some content words, others a conjunction, a pronoun, etc.), and asked the child whether each was a word or not, and why.

-- In a third experiment, children were asked to give examples of words with certain properties (a *long* word, a *short* word and a *difficult* word), and were encouraged to justify their answers. Length and difficulty were chosen because they can be attached to what is signified as well as to the linguistic signifier without any kind of metaphoric extension. It was hoped that children's answers to these questions would tell us something about the link they had established between elements of language and objects in the outside world.

-- In yet another experiment, children were asked to *invent* a word and to produce an utterance containing it. Despite the non-communicative nature of this task, it seemed interesting nevertheless: a different kind of manipulation of the link between language and what language represents is implied. The children would hopefully show us something of their ideas about the creation of new words, and, through the use of these words in sentences, something about their form, meaning and function.

-- A fifth and final experiment ran as follows: the experimenter uttered a sentence and asked the child to count the words. According to the child's response ("one, two ... n words") he was asked "What were those n words?" Several different sentences types were presented in order to see what units the child would use when asked to analyze well-formed sentences in this way. Although this task is not one of direct segmentation, children's answers may tell us something about how they break up sentences into smaller units. The questions were phrased in a way that allowed the children to express the idea that words and other elements are what make utterances up.

A total of 163 children aged four to twelve years took part in the experiments. All were attending nursery or primary schools in the Canton of Geneva, Switzerland. Each child took part in at least two of the experiments, and many took part in all five. The order was varied from subject to subject. During the discussion the experimenters carefully avoided using terms that were highly suggestive or too difficult. Children often used the metalinguistic terms *word* and *say* spontaneously, and other terms were used in the discussion if they were first mentioned by the child (e.g., *speak*, *talk*, *mean*, and, for older children, *verb*, *adjective*, *sentence*, etc.). Each experiment elicited responses that were classifiable by age; a general trend in response types was noted over the set of experiments. The following examples, typical of the successive response patterns observed, will give some idea of this general developmental trend. Rather than presenting all the response types observed for each age group, we will describe the main response patterns, and simply state the ages at which they are preponderant.

The youngest subjects (four- and five-year olds) show failure to differentiate conceptually between words and things (or actions): words are placed on the same level as extralinguistic elements and are thought to share some of the latter's features.

For example, from a child aged 5;4: "Un mot c'est quelque chose qui existe, c'est quelque chose de vrai, parce qu'on voit ce que c'est" ('A word is something that exists, that is true, because one can see it'), and "Fraise c'est un mot, parce que ça pousse dans le jardin" ('Strawberry is

58

a word, because it grows in the garden'). At this age, *train* is often proposed as a long word. Other examples for *long* words which show that the child takes into consideration the objects referred to are: (4;1) "Armoire, parce qu'il y a beaucoup d'affaires dedans" ('Cupboard, because there's a lot of stuff in it'); (4;9) "Chaise, les barres pour tenir, elles sont longues" ('Chair, because it has long legs to hold it up'). And, correspondingly for a *short* word the following responses are representative: (5;4) "Etoile" ('Star'); "Un oeil, parce qu'il est petit" ('An eye, because it's small'); (4;7) "Primevère" ('Primrose').

Similarly, when asked to produce a *difficult* word, lack of differentiation is again evident, as for example: (4;5) "Quelqu'un qui enlève la clef, parce que c'est difficile" ('Somebody who takes the key out, because it's difficult'); (5;11) "Il range, parce qu'il doit ranger tous les jouets" ('He tidies up, because he has to tidy up all the toys'). However, for difficult words, even slightly more advanced children (six and above) already attribute the property only to the word, and justify their choice by saying that the word is difficult to say, to write, or to remember.

Some subjects from four to six define words as the act of speaking itself: "Un mot, c'est quand on parle, quand on dit quelque chose" ('A word is when you speak, when you say something'). This approach often leads children to given entire sentences as examples of a single word: "On peut parler avec un mot, un peut dire *viens ici* " ('You can speak with a word, you can say *come here*'). Asked "tell me a word," these children give sentences such as: "Ma soeur voulait jouer avec moi" ('My sister wanted to play with me'). For a long word, they often give two or more sentences in contrast to a single sentence for a short word. For example:

*A long word:* (5;11) "Il s'en va et puis il monte sur la voiture ('He goes away and then he climbs onto the car'). *A short word:* "Il s'en va." Experimenter: "Why is that a short word?" "Parce qu'il s'en va seulement" ('He goes away') ...('Because he only goes away').

*A long word:* (6;9) "J'entre dans la maison pour quitter mes chaussures" ('I am going into the house to take off my shoes'). "Why is that a long word?" "C'est un mot long parce que j'ai dit deux choses à la fois" ('It's a long word because I said two things at once'). *A short word:* "Je vais là" ('I'm going there'). "Why is that a short word?" "C'est un petit mot parce qu'il y a qu'un mot, non, il y a qu'une chose" ('It's a small word because there is only one word, no, only one thing'). In these examples it is the length of the message as well as the amount of content that determines the length of what the child considers to be a word.

At a slightly higher level, words become still more differentiated from the physical reality they represent. Mainly at the ages of six and seven, words are seen as labels that correspond to things, and as such, have an independent existence. They have become composite units: "Un mot, c'est des lettres" ('A word, well, it's letters'). This definition is often accompanied by restrictions such as: "Il faut assez de lettres pour faire un mot, pas trop, ni trop peu" ('You need enough letters to make a word, not too many and not too few'). Such restrictions make it possible for children at this level to exclude articles and other functors from the class of words. For example, *le* in *le garçon*: "C'est pas un mot, parce qu'il n'a pas beaucoup de lettres" ('... isn't a word, because there aren't enough letters'). But when the experimenter then says: "And *âne*, is that a word or not?" the children use the other criterion, and explain that "c'est un mot, parce que les ânes, ça existe" ('It is a word, because donkeys exist'). The reference to reality overrides the criterion of a certain number of letters.

Words are considered to be larger meaningful units from the age of seven onwards but not before. "Un mot, c'est un bout de l'histoire" ('A word is a bit of a story'), or "Un mot c'est tout seul, ça ne raconte pas quelque chose, on a une phrase, il y a des mots dedans" ('A word says nothing all alone, it's not a sentence, when you say a sentence there are words in it'). These are typical responses at those ages. From the age of eight onwards, children refer not only to the analysis of the spoken chain, but also to it's synthesis: "C'est un mot qu'on peut servir dans les phrases" ('Words are used to make sentences with').

When asked to produce *long* or *short* words, children of seven and above appear to situate word-length at the level of the graphic representation of the signifier. A word is long because it has lots of letters. But before the length of words becomes completely detached from the size of their referents, the children invent interesting compromises between the length of the word and some quantitative property of the referent. For example:

(i) *A long word:* (6;10) "Un crocodile" ('A crocodile'), and a *short word:* "Le chat" ('The cat').

*A long word:* (6;0) "Une machine à écrire" ('a typewriter'). "Why is that long?" "C'est long parce qu'elle a beaucoup de lettres" ('It's long because it has a lot of letters').

   *A long word:* (6;8) "Journal" ('Newspaper'). "Why is that long?" "Il
y a beaucoup de choses écrites" ('There's a lot of written stuff').

   By the age of seven or eight, children also attempt to define words by
using certain grammatical terms without making it clear that these terms
refer to subsets: "Words are adjectives, nouns and verbs." The first truly
grammatical criteria are given in the form of rules that govern the links
between words; they appear around the age of eight and are at first only
applied to nouns. At this age, the definitions given for *word* are more or
less exhaustive and correct definitions of the French term *nom*. In French,
the word *nom* is used both as a grammatical term, equivalent to *substantive*
or *noun*, and in non-technical discourse, where it means *name*. The children
include in their definitions characteristics of both meanings of *nom*: (again
to paraphrase in English) "A word is the name of a person, an animal, a thing,
a flower, it's also when you can put *the* in front, it's when you can put it
in the singular or the plural." This approach is characterized by the ex-
clusion of all words other than nouns, which seem to be the prototype of
words that can be isolated, and for which children can give some kind of
semantic description as well as some correct and precise construction rules.

   Meaning is appealed to in children's answers as of age seven, but at
first only to reject a counter-example when a nonsense word is reported.
Nonsense words are not words, "parce qu'ils ne veulent rien dire" ('because
they don't mean anything'), though they are thought of as being "fait de
lettres" ('made up of letters'). Meaning is systematically mentioned as
part of the definition only at age ten or later: "Un mot c'est fait de lettres
et ça veut dire quelque chose" ('A word is made up of letters and it means
something'). Furthermore, meaning is attached to all types of words but in
different degrees: (11;7) "*Table*, ça veut dire un meuble. *La* c'est un article, ça accompagne table, ça veut dire si c'est au singulier" ('*Table* means
a piece of furniture, *the* is an article that goes with table, it means it's
in the singular'). At this level, the definitions are completed by reference
to the sentence as well as to the formal categories and to the grammatical
rules that link them. Meaning, substance and form of the signifiers and the
fact that they are governed by a rule-system are all taken into account.
The definitions are at the same time more general and more explicitly linguistic, and no longer focus only on some particular aspect of words.

   Intriguingly, one aspect of words--their graphic substance--seems to

occupy a very special place in children's thinking. A large number of subjects between the ages of six and twelve (56 in all) mentioned the letters
words are made up of, while only six subjects referred to their phonic substance. We may well wonder why letters and not sounds are mentioned so often. Is the children's choice determined by the influence of school where
they have to concentrate on written or printed words? Or are other factors
also at work? Some speculations may be in order. The written word, because
of its permanent and objective qualities, lends itself better to contempla-
tion and study than speech. Speech is dependent on the speaker and the
hearer, and is limited to the here and now of the act of speaking. Put down
in black and white, words may well acquire a more substantial identity of
their own. Moreover, in our alphabetic writing system, words are separated
by blanks, whereas in the continuous spoken chain no such segmentation can
generally be heard; thinking about words as made up of smaller elements may
therefore be easier when one imagines them in written form. However, the
experimental data do not allow any firm interpretation on this point (see
also Lundberg in this volume).

The results of the experiment on the counting of words in spoken sentences confirm the general developmental trend already described. At the
youngest ages children appear to focus on the scene evoked by the sentence,
i.e., on the tangible elements of reality it represents. (4;1) *Six enfants
jouent:* ('Six children are playing') "How many words?" "Six" "What are
those six words?" "Moi, mon petit frère, et Christiane, Anne, Jean, etc."
('Me my little brother, and Christiane, Anne, Jean, etc.') *Le garçon lave le
camion:* ('The boy washes the truck'). "How many words?" "Un" ('One').
"What is the one word?" "Parce qu'il lave le *camion*". ('Because he washes
the *truck*'). At the same level, or slightly later, children may also focus
on the rhythmic and syllabic aspect of the spoken sentence. For *six enfants
jouent,* "How many words?" "Sixenfantsjouent: (si/zan/fan/jou/).. quatre."
('four') (The child lifts a finger for each accented syllable).

The same children who give these answers are often already capable of
carrying out a very general segmentation of the sentence into topic and comment. *Six enfants jouent:* "How many words?" "Deux" ('Two'). "What are
they?" "Il y a six enfants et jouent" ('There's six children, and that they
play'). *Le garçon lave le camion:* "How many words?" "Deux" ('Two') "What
are they?" "*Le garçon et il lave le camion*" ('The boy and he washes the
truck'). Gradually, another kind of cut is introduced and the children count

what may be called the privileged constituents: *Le garçon lave le camion* –
"How many words?" "Trois" ('Three') "What are they?" "Le garçon, lave,
et le camion" ('The boy, washes and the truck'). Articles and other functors
are not counted separately until a later age; depending on the particular
sentence presented they are sometimes included at the age of seven, but they
are included systematically, for all sentences, only from about the age of
11. Responses like "Garçon, lave, camion, trois" ('Boy, washes, truck,
three') correspond to the restrictive conception of words at the age noted
for the other experiments.

When all the responses of one child are compared over experiments in
which he took part, they may seem heterogeneous though they are not, in fact,
contradictory. The lack of homogeneity is partly due to the fact that each
experiment sought to explore different aspects of words. Length, for example,
is an objective physical property of words and of objects, whereas difficulty
is a very different, more subjective property, often applied to the performance of actions. It appears that well before children can dissociate the
"length" of an object or an action from the word that expresses it, they
are capable of making a distinction concerning "difficulty". Their responses
then consist of words that are difficult to pronounce, to read or to remember. Lack of homogeneity may also derive from the context in which the word
is presented in the experiment. For example, articles or auxiliaries may be
accepted as words when presented in isolation, and at the same time be ne-
glected in the word-count experiment, where they remain attached to the noun
or verb they accompany.

The lack of homogeneity is observed above all in the intermediate age
group. The four- and five-year olds concentrate, in all the experiments,
on the real situation that the words or sentences describe. The oldest
subjects, on the other hand, have elaborated a clear concept of what words
are, and can make it explicit in the various experimental contexts. Children in the intermediate age group, however, have not yet worked out how the
different criteria can be combined, and see nothing contradictory in, for
example, accepting *dormir* as a word, "parce qu'on dort dans un lit" ('because
you sleep in a bed'), and rejecting *la* "parce qu'il n'y a pas assez de lettres" ('because there aren't enough letters') or accepting it, "parce qu'on
l'écrit" ('because you write it').

Beyond the specific results of each experiment a general developmental
trend in metalinguistic thinking appears in the sense of a reflection on the

nature of language itself: whether in the characterization of words, or in the judgments and production of words, or in the segmentation of sentences, metalinguistic thinking first focuses either on the elements of reality designated by the linguistic units or on the speaker, who is himself real but possesses the capacity of referring in speech to all aspects of reality, himself included. Becoming aware of what goes on when one names objects or actions, utters a command, or when one reads or writes, the child detaches signifiers from their signifieds and starts to reflect on the correspondence between words and things (in the larger sense of the word). At this level, words become composite elements that have status proper in their graphic or phonic substance. Simultaneously, the child becomes aware of the possibility of analyzing his own (or others') utterances into smaller units, and thereby adds another aspect to his concept of words: they are constituent elements of larger units. At the highest level observed in our experiments, the children were also able to make explicit the link of words as signifiers to what they signify, and the links that join words together into groups of units in general discourse.

*ACKNOWLEDGEMENT*

The research on which this paper is based was carried out with the help of the Fonds National de la Recherche Scientifique Suisse, Grant No. 1.190 - 0.75 and 1.527 - 0.77.

*FOOTNOTE*

1. A comparison between children's metalinguistic reflection and the history of linguistics may be both justified and profitable, as research on metalinguistic activity can also be viewed within the framework of the history of science (Papandropoulou & Sinclair, 1974). Such considerations may lead to better understanding of the link between the study of language on two seemingly distant planes, that of the linguist and that of the psychologist.

*REFERENCES*

Berthoud-Papandropoulou, I. *La réflexion metalinguistique chez l'enfant.* Doctoral thesis, University of Geneva, 1976.

Hockett, C.F., & Altmann, S.A. A note on design features. In: T.A. Sebeok (ed.), *Animal communication.* Bloomington: Indiana University Press,

1968.

Kaper, W. *Kindersprachforschung mit Hilfe des Kindes: einige Erscheinungen der kindlichen Spracherwerbung erläutert im Lichte des vom Kinde gezeigten Interesse für Sprachliches.* Groningen: Wolters, 1959.

Papandropoulou, I., & Sinclair, H. What is a word? Experimental study of children's ideas on grammar. *Human Development*, 1974, 17, 241-258.

# Children's Awareness of Language, with Emphasis on Sound Systems

Charles Read

University of Wisconsin, Madison, WI 53706, USA

## What is Metalinguistic Awareness?

For a start let us suppose that metalinguistic awareness includes the ability to think about language and comment on it. A linguist or a language-learner studying a grammar of an unfamiliar language is acquiring one type of metalinguistic awareness. But in the more interesting case, the fluent speaker of a language typically has some awareness of the language that he or she speaks. We might say that a speaker exhibits metalinguistic awareness when that speaker is attending to some part of what he or she knows about the language, and also knows that he or she possesses that knowledge.

This distinction, between knowing something and knowing that one knows it, is one that linguists commonly make. For example, a person who can form regular English plurals, even of unfamiliar words, must know that words ending in /p,t,k/ take plurals in -/s/, while those ending in /b,d,g/ take plurals in -/z/; but typically it is only those people with a degree of linguistic training who know that they know such a principle. We must distinguish further between awareness of linguistic structure, as in the case of plural formation, and awareness of linguistic performance. For example, there is an important difference between the teacher who mystifies his or her students by taking too much for granted but then recognizes the problem, and a teacher who commits the same error in blissful ignorance.

These distinctions--between awareness of one's native language and awareness in formal language study, and between awareness of structure and awareness of performance--help to account for the great diversity of examples

given in the papers in this volume. The contributors have discussed tasks
ranging from the anticipation of conversational turns to formulation of explicit linguistic rules, all under the general heading of linguistic awareness. While I doubt that linguistic awareness in any of these forms is
strictly necessary to the basic acquisition of a native language, I believe
that the analytic and judgmental functions that have been illustrated do
have great practical significance. The performances of adapting, manipulating, segmenting, correcting, and judging language seem to play an important
role in at least three processes: learning to read and write, learning a
non-native language, and responding to social expectations. In short, they
have a great deal to do with using language effectively under varied circumstances. Whether they are conscious or can easily be brought to consciousness
appears to be of secondary importance. I shall return to the role of judgment
and analysis in language use, but for now wish to focus on what one knows
about the structure, particularly the phonological structure, of one's own
language.

Perhaps the first question about awareness of linguistic structure to
be asked is, "Why should we be interested in it?" If contemplating our language is not necessary for speaking and listening under ordinary circumstances, might we doom our investigations to quaintness by studying a non-
essential and even unusual kind of behavior? Certainly some such view has
been standard in psychology and linguistics since the decline of introspec-
tionism.

### Its Usefulness to Psycholinguistics

At the first level, the usefulness of evidence from metalinguistic
awareness is that it adds judgments and corrections to the usual range of
evidence to be found in children's language, namely evidence from production
and comprehension. This move enriches our data considerably, making it
possible (a) to check the validity of inferences drawn from production and
comprehension, (b) to extend our evidence to sentences which might not occur
in spontaneous production or which might be difficult to present in a comprehension task because they are not easily depicted or acted out, for example, and (c) to deepen our understanding of why language development proceeds as it does, by providing evidence that cannot be obtained from production or comprehension, such as judgments of structural parallelism. As an
example, consider the striking +.77 correlation between correction of object-

verb word order in English imperatives and accurate interpretation of reversible passives found by de Villiers and de Villiers among four-year old children (1974, p. 18). By tying together two aspects of language development which need not have been related, this observation suggests that a concept of expected word order, applying across sentence types, may develop in children at approximately age four.

*Awareness of Structure as Opposed to Awareness of Rules*

We have, I think, an inclination to reason in the following way:

Question: Who has metalinguistic awareness in its purest form?
Answer  : Linguists, of course.
Question: What is it that linguists most notably possess?
Answer  : A knowledge of the rules (generalizations) which describe our language(s).

From this little dialogue, we are tempted to conclude that it is just such awareness of rules which we want to observe in children. Such a focus is suggested, for example, in the first pages of Gleitman, Gleitman, and Shipley (1972), where they use the analogy of a chess player who can articulate the rules of the game, or perhaps the rules that he or she follows in playing the game. It is instructive to note, however, that what Gleitman, Gleitman, and Shipley actually chose to study were judgments of grammaticality or deviance. That is, they studied comments on *sentences*, rather than on rules, processes, or generalizations. I believe their decision was correct, and that in general we will have better success in eliciting judgments of structures and structural relationships, such as grammaticality, parallelism, and ambiguity, than in eliciting comments on linguistic rules or generalizations. Although both structural judgments and knowledge of rules are generally included under the heading of linguistic awareness, their subjects are quite different: on the one hand, the properties of one sentence at a time, on the other, generalizations across many different sentences. It is no surprise that children, like adults, are better at commenting on the former than the latter. Even to begin to formulate generalizations about linguistic processes, one has to decide what sentences have the relevant properties, whereas one can notice the ambiguity of a sentence without having the least awareness of what generalizations apply. There is also a growing conviction that what is psychologically real, or at least most accessible, to the language user are the levels of structure that a grammar describes, not the rules which link one level to another (cf. Fodor, Bever, & Garrett, 1974)

### Awareness of Phonology

From the foregoing comments on linguistic awareness in general, let us consider the ways in which children might demonstrate an awareness of the sound system of their language in particular. Under the somewhat misleading heading of phonology, psycholinguists have investigated three aspects of performance, each with its characteristic paradigm. These have been:

(i)  at the phonetic level, production of phones;

(ii)  at the phonemic level, discrimination of members of distinct phonemes;

(iii)  at the phonological level, judgments of the acceptability of sequences.

Of these it is the phonological sequence studies which seem to require some metalinguistic awareness and which have a parallel in judgments of syntactic acceptability. I believe there is much more to be learned from such judgments. Both Messer (1967) and Menyuk (1968) have dealt with constraints on initial consonant clusters in English, and Messer (1967) and Finnegan (1976) also with final clusters. But relatively few phonological constraints have been studied, even in a well-worked language such as English. For example, neither the familiar principle of nasal assimilation nor any of the constraints on polysyllabic forms including stress assignment, have been studied with acceptability judgments. A further consequence is that we really have no information about the relative ages at which different kinds of phonological constraints become accessible, and yet know that listeners can use these constraints (for example, in identifying words under noisy conditions).

In addition to further study of children's ability to distinguish possible sequences, however, there are experimental tasks which have so far been little used, and might help us to understand new facets of children's phonological awareness. These include judgments of relative similarity among members of distinct phonemes, which may lead to a better understanding of how children categorize or cluster their language phonemes (cf. Read, 1975, chapters, 3, 5, and 6). Another kind of information is to be gained from children's judgments of rhyme or their ability to re-sequence phones, as in

the formation of Pig Latin and other "secret" languages. Such studies may
shed light on the important question of how well children can identify the
individual segments of words. Savin (1972), for example, makes some rather
remarkable observations about the extent of individual variation in this task,
but admits to having little hard evidence. Also interesting is the possibility of asking children to locate certain sounds within a printed representation of a word, as in Read (1975, pp. 108-112).

## Segmentation into Phonemes

Before suggesting other topics for investigations, however, I would like
to consider three questions about children's ability, or inability, to segment utterances into their constituent phonemes: (1) Why is there a contrast
between one study and most of the others on children's segmentation at age
five or six? (2) What is the theoretical significance of the role of training in these studies? (3) Is the difficulty which some children have in
segmentation really related to phonetic complexities?

Most recent studies of children's ability to count or manipulate the
segments of a word have concluded that the task is difficult for most children younger than age six or seven. Bruce (1964) reported that children
with a mental age greater than seven years were able to identify the word
that would be left if, for example, one removed the first sound from *pink*.
McNinch (1975) found that children in second grade, but not first graders,
could select a picture of a *map* as a word which has the same sounds as *pam*,
but in a different order. Liberman, Shankweiler, Liberman, Fowler, and
Fischer (1977) trained children to tap on a table to indicate the number
of sounds in a word; only 17 percent of five-year olds but 70 percent of
six-year olds succeeded. From such varied experimental methods with rather
uniform conclusions, one might well infer that it is not until age six or
seven that children typically develop (or acquire) the ability to segment
words into the sounds which make them up. But Zhurova (1973) taught children to pronounce the first sound of a word in a game situation in which the
experimenter gave examples like *b-b-b-bear*. Zhurova reports that the three-
and four-year olds succeeded as long as the model was provided, and that
four- and five-year olds continued to perform successfully even in the absence of the model.

How are we to reconcile these Russian and English results, considering
that all four tasks do seem to involve segmentation in some sense? It is

entirely likely that one difference has to do with the specific task: pronunciation versus counting or manipulation. Also, it may be pertinent that Zhurova's task focused on the *first* sound in the word, while McNinch's and Liberman et al.'s involved all of the segments, and Bruce tested segments in three different positions within words. These are questions which remain to be resolved.

*The Problem of Training*

The success of Zhurova's teaching, however, suggests a larger conceptual problem. With three- to five-year olds, Zhurova evidently succeeded in eliciting a segmented pronunciation which many children were unable to provide at the outset of the experiment. Are we to conclude that the experimental training changed the child's level of linguistic awareness? If we take "awareness" to mean *focusing one's attention on something that one knows*, it seems that the children's awareness changed, and indeed that it changes generally: all of the experiments on segmentation involved some training; Zhurova's was merely the most successful in that respect.

The often substantial effect of training on children's judgments of alliteration, rhyme, segmentation, or grammaticality seems to pose an unpleasant dilemma. Surely these judgments are legitimate examples of linguistic awareness, but just as surely we do not want to study a phenomenon which may change right before our eyes, as we conduct even a modest experiment. If there is to be any point in describing linguistic awareness, must it not stand still for at least a week or two?

My own view is that linguistic awareness is indeed unstable, but that our response to my last question must be "No." As work in nuclear physics reminds us, phenomena which we cannot observe without affecting them may nonetheless prove to be extremely interesting.[1] It should come as no surprise that linguistic awareness may be changed by training, or even merely by presenting examples in the course of an experiment. Some aspects of language use or structure do seem to lie just below the threshold of consciousness. For instance, perhaps most people are unaware that there is a social constraint on standing too close to one's listener, and yet merely being confronted with a violation of that constraint may make a listener aware of it. Similarly with structure: while most of the structure of our language resists inspection, still we often find ourselves noticing that we have inadvertently created an alliterative phrase, or one with a catchy rhythm, one that "sounds wrong," one with a double meaning, and so on. Included in our linguistic awareness are judgments which pass in and out of consciousness, depending on the momentary stimulation. In children, such judgments may often be elicited or changed with just a minimum of exemplification or training. This instability is a decided inconvenience for the experimenter, but it does not render the judgments uninteresting. The experimenter must estimate the effect of (intended or unintended) experimental training, must be careful in generalizing to the non-experimental context, and must be extremely cautious in concluding that children at a given stage *cannot* be brought to make a particular judgment. In some cases we cannot directly observe but only estimate the condition of the child before our interview or experiment, but these cautions need not diminish our interest in the subject matter at all.

These considerations do suggest the need to distinguish between one's *awareness* of linguistic structure and the *accessibility* of a structure. George Miller urged such a distinction in commenting on a conference which, like this one, had a great deal to say about linguistic awareness (1972, p. 378). An accessible structure is one that may be brought to awareness, even if the subject has never been aware of it previously, and even if only training of a special kind can make the subject aware of it. Concerning the children who succeeded in segmentation only after training, we can then say that the accessibility of segments for them probably did not change, although their awareness of segments did, at least momentarily.

This distinction seems natural enough and it has been implied in much of the experimental work in the field, but it has an interesting consequence. It suggests that knowledge which is accessible might be reflected in behavior in various ways, not only through being brought to awareness. Might not a knowledge of grammaticality, ambiguity, paraphrase, structural parallelism, segmentation, categorization, similarity, and other judgments that we have grouped under "awareness" be reflected in various kinds of behavior, even without the subject's being *aware* of what he or she knows? In this broader sense, a paradigm such as the perceptual "click" experiments (Fodor, Bever, & Garrett, 1974) might belong in the same class as explicit judgments of syntactic structure, even though in the "click" experiments, the subject is not aware that the location he or she reports for the click has anything to do with syntactic structure.

In short, as we recognize the continuous variability of linguistic awareness, we can no longer be certain of the distinction between conscious

judgments and any other behavior which reveals a knowledge of linguistic
structure. As evidence, a speaker's awareness of what he or she knows enjoys
no privileged, nor for that matter disadvantaged, status. There is a con-
tinuum from linguistic knowledge of which we are spontaneously aware to that
which remains totally out of awareness. The fact that awareness changes with
children's level of development, and with training, tends to blur any distinction between knowledge which we are aware of and that which we are not
aware of.

*Why is Segmentation Difficult?*

The third problem with children's ability to identify phonetic segments
is that the task appears to be relatively difficult, compared, for instance,
to identifying the syllables of an utterance. Liberman et al. (1977) related
this difficulty to the fact that in acoustic terms the segment simply does
not exist. Our perception that *cat* has three segments is one that we impose upon the signal. The physical signal is continuous, with no unambiguous
gaps or transitions between segments. In fact, within a syllable and even
across longer streches of speech, one segment influences its neighbors, so
that in *bag* the vowel contains evidence of both the [b] and the [g].

This explanation for the difficulty of segmentation has been coupled
with the speculation, based on the history of writing systems, that identifying words and syllables ought to be relatively easier (Liberman, et al.,
1977, p.209). However, history is misleading on this point. Identifying
words turns out to be a quite difficult task, and one which also develops
only in the early school years. There is a substantial literature, some of
it directly associated with reading instruction, documenting the difficulty
with which children identify words (Karpova, 1966; Downing and Oliver, 1973-
74; Johns, 1976, among others).

The word and the phonetic segment have some characteristics in common;
both appear to be units accessible to the speaker (or at least to the reader!)
and yet both have defied linguistic definition. Linguists have offered some
rather lame definitions of the word, such as Hockett's "any segment of a
sentence bounded by successive points at which pausing is possible" (1958,
p. 167). Basically, however, it is a source of minor embarrassment that the
word seems to be a significant unit to speakers and yet does not play a very
important role in linguistic description. The situation is even worse for
the phonetic segment; it is real to speakers, but linguists usually don't

even try to define it. It is clearly a linguistic prime. The syllable, too, has resisted satisfactory delimitation, but there is a simple acoustic criterion for the number of syllables, if not for their boundaries.

Although words are difficult for children to identify, they do not embody the phonetic coding, or overlapping, characteristic of segments. Words are not separated by pauses or other obvious boundaries, it is true, but words do not affect their neighbors phonetically to the extent that segments do. While phonetic overlapping or encoding may help to account for the elusiveness of the segment, some other explanation will have to be given for the word.

One property that words and segments have in common is that both are highly abstract from the acoustic signal; there is no simple physical basis for recognizing either a word or a segment. The syllable peak, on the other hand, has a physical correlate in relative amplitude. What children's difficulty with these units suggests to me is that children begin their analysis quite close to the phonetic level. As I have argued elsewhere, they seek out relationships among phones (Read, 1975), but these are relationships in sound, not function or distribution. From this point of view, it is not surprising that the word and the segment are relatively difficult for children to isolate: as a class, words have little phonetic resemblance to each other, and the same is true of phonetic segments. Hayes and Clark (1970) report experiments in which adults isolated word-like units in artificial speech, apparently using a distributional analysis. It is unclear whether children analyze real language in the same way.

One final observation about the segmentation problem is that the appearance of segmentation ability near age six or seven is decidedly suspicious; one wonders whether it is reading instruction and reading experience which brings these units to awareness. Bruce's (1964) results varied significantly with the type of reading instruction that the subjects had experienced, for instance. Johns (1976) found that it was only the children in second or third grade (aged 8;1 and up) who were able to distinguish words from other units at least 80 percent of the time. Clearly, the problem of whether this awareness depends on instruction, or successful instruction depends on this awareness, is amenable to cross-linguistic and cross-cultural study. I do not know of any study which has succeeded in clarifying this relationship.

*Children's Categorization of Vowels*

Let us now return to examining research designs which appear promising

but have been little used. A set of experiments I conducted to test whether
children categorize certain English vowels is pertinent here, because it
illustrates both the possibility of eliciting judgments of phonetic relationships from children and the importance of the specific form in which the task
is presented. Eliciting explicit judgments of relatedness contrasts with
other paradigms, more commonly used with adults, involving the recognition
or discrimination of speech sounds under noisy conditions, or the study of
errors in short-term memory (Campbell, 1970; Miller & Nicely, 1955; Wickelgren, 1965, 1966; etc.). Each procedure has its advantages; one advantage
of the explicit judgments, if they can be elicited reliably, is that they
may be more closely related to higher level judgments involved in reading
and writing, and less influenced by the fact that specific noise frequencies
mask particular characteristics of speech sounds (Singh, 1971).

My reason for studying children's categorization of vowels was the observation that there are certain regularities in spellings invented by preschool children. In my sample of these early spellings, children tended to
spell the vowel of *bed* with an *A*, for example. In fact, this particular
spelling occurred 50 percent of the time in writing of children younger
than age six, more frequently than the standard spelling, *E* (37%), and far
more frequently than any other non-standard spelling. Except for two spellings for which there are particular reasons, all other non-standard spellings of this vowel occurred 1 percent of the time or less (Read, 1975, pp.
37-41).

Attempting to account for this disproportionately frequent non-standard
spelling, I hypothesized that children may recognize a relationship between
/ɛ/ and /e/, the vowel of *bed* and *bayed*. Phoneticians have considered these
to be related in that they are the two mid front vowels in English, distinguished as "short" and "long," "lax" and "tense," or "monophthongal" and
"diphthongal," respectively. However, it is interesting to note that there
are no strictly *phonetic* reasons for pairing these two vowels rather than
two others, such as /ɛ/ and /I/. One reason that they are considered to be
related is that they enter into phonological alternations in various languages
but of course my English-speaking five-year old spellers knew very little
about such alternations. It would be interesting if the non-standard spellings reflected a categorization which has long been implied in discussions
of phonetics and phonology. This explanation gains some strength from the
fact that children likewise tended to spell other tense-lax pairs of vowels

alike. There are other conceivable explanations, of course, but most of
them appear to be ruled out by one consideration or another.

In order to test whether a categorization of vowels might be the basis
for the spelling, we tried to design an experiment in which children would
be induced to indicate whether they found certain pairs of vowels more similar than others. Our first attempt used an X:A, B paradigm similar to that
used by Singh (1971) to study relationships among consonants as judged by
adults, indeed by phonetics students. In this experiment, we presented a
nonce word, such as /pek/ (X) and asked children whether it was more like
*peck* (A) or *peek* (B). There were two such patterns, testing whether children might find /ɛ/ and /e/ to be more closely related than another pairing,
with three instances of each pattern. In deference to the age of our subjects, we introduced hand puppets into the procedure.

The experiment and its results are described in Read (1973, pp. 17-26).
My methodological observation is simply that it didn't work at all with five-
year olds, and it failed to elicit large numbers of reliable, consistent
judgments from six-year olds. Only from seven-year olds did we receive
statistically significant preferences, and then on only one pattern of this
type. Substantial numbers of children were inconsistent in their judgments.
The appearance of the ability to make these judgments consistently at age
seven resembled the results of the segmentation studies and suggested an
influence of schooling. We tried various methods of eliciting consistent
judgments from younger children, for it seemed important to discover whether
children recognized relationships among speech sounds *before* they learned
to read and write. One effort was to reduce the role of short-term memory
in the X:A, B paradigm, where the subject has to remember X while comparing
it to both A and B. In this attempt, we tried variants such as XA:XB presentations.

More than a year after our first studies, a relatively small change in
our methodology considerably enhanced the success of this task with young
children. We introduced a hand puppet named Ed and announced that Ed liked
to find words that sound like *Ed*, such as *Ted, Jed, fled,* and *sled.* We then
trained children, to the extent that they needed any training, to choose from
a pair of the words the one that rhymes with ("sounds like") *Ed:* Would Ed
like *bed* or *bead*? Would Ed like *food* or *fed*? Among the kindergarten children we tested, 82 percent achieved the criterion of identifying at least
five out of six rhyming words in six consecutive trials, although 25 percent

of them required more than six trials to do so. To those who met this criterion, we then said, "Now I'll tell you some words that don't sound exactly like *Ed*, and you ... tell me which one Ed would like. Would Ed like *aid* or *owed*? ... Would Ed like *showed* or *shade*?" This experiment and its sequels are described in Read (1975, pp. 120-126). The methodological point is that this change in the task made it accessible to most kindergarten children, two years younger than those from whom our earlier paradigm elicited consistent judgments. It also elicited much more spontaneous and confident participation.

In response to the first question, "Would Ed like *aid* or *owed*?", 15 of 20 kindergarten children chose *aid*, a result which would occur by chance alone only twice in one hundred times (cumulative binomial probability, one-tailed). This is a conservative way of looking at the data; it eliminates any assumptions about population distributions or independence of the six trials of this type. Other ways of looking at the results also show that the children paired /ɛ/ with /e/, rather than with /o/, to a significant degree. We then went on to examine other relationships among vowels. We found that children tend to pair /ɛ/ with /æ/, rather than with /ɔ/ (p=.01), for instance, and that the /ɛ/ - /æ/ relationship is stronger than that between /ɛ/ and /e/ (p=.09). In other words, children do not regard the vowels of *bait, bet,* and *bat* as simply three distinct vowels of English. Rather, they recognize that these vowels are phonetically related, and they judge that these are more closely related than other possible pairs of vowels.

This result suggests what I believe is the right explanation for the spelling: among the front vowels, /ɛ/ is flanked by two vowels that children know are related to the letter *A*; /e/ is the name of that letter, and /æ/ is spelled *A*, particularly in some frequent and early-learned words, like *man*. Not knowing how to spell /ɛ/, but knowing that it is related to these vowels, children conclude that *A* is the most likely spelling. This reasoning illustrates how close children are to the level of sounds when they first encounter reading and writing. Their judgments are perfectly reasonable and phonetically correct; they would even be right, were it not for the fact that English spelling no longer closely reflects relationships at the phonetic level. Beyond accounting for this invented spelling, these experiments suggested a "map" of English front vowels as seen by children, in which the strongest relationships hold between vowels that are alike in diphthongization, and the weaker, but noticeable, similarities hold between vowels that are alike in height (Read, 1975, p. 127).

In this case, the methodological changes that mattered were that we

-- embedded the experimental task in a "story" that made some sense, rather than asking for a judgment out of context;

-- gave children practice on a simpler task, finding a true rhyme, before asking them to make a more difficult judgment; and

-- gave children a consistent "target," /ɛd/, for their judgments, rather than giving them a new target on each trial.

These changes might have been relatively minor with adults, but they turned out to be quite important in studying children.

The substantive result was the discovery that one can elicit judgments of phonetic similarities from children even before they learn to read and write, considerably earlier than we might have expected from the use of other techniques, both on categorization and on segmentation. We need not limit ourselves to discrimination paradigms in order to study phonetic relationships in children. This constitutes progress, because discrimination tasks produce little information (low error rates) unless one undertakes a very large study (Graham & House, 1971) or introduces masking noise.

An interesting observation about judgments of phonetic similarity is that they are not required for successful speaking and understanding under ordinary circumstances. All that is required for speaking and listening is the ability to distinguish the members of discrete phonemes; the phonemes need not be related to each other. In their judgments of phonetic similarity, children show that their language development includes knowledge which is not strictly necessary for successful communication. Also, because judgments of phonetic similarity are not required for communication, we present evidence of them only very indirectly in our speech, as in puns, off-rhymes, and phonological alternations. Linguistic knowledge which cannot be derived by simple inference from plentiful examples in everyday speech may be particularly revealing of how acquisition proceeds. Part of the importance of research on linguistic awareness is the possibility of identifying that kind of knowledge, and perhaps learning something about why it is acquired.

These judgments of vowel similarity display some other interesting properties, which may extend to other kinds of linguistic awareness, as well. For one thing, these are judgments which change, and from a phonetic point of view deteriorate, in adulthood. Adults typically judge that the vowels of *bet* and *beet*, *bat* and *bait*, and *bite* and *bit* are the related pairs because of the similarities in spelling, even though these pairs are not especially similar phonetically. Familiarity with the written form eventually drives the phonetic relationships out of our awareness, it seems. Beginning students of phonetics usually have to work to acquire (or re-acquire) the judgments which the kindergarten children can make.

Also, the children's spelling illustrates some limitations of their linguistic awareness. Their grouping of vowels leads to a substantial degree of homography: *Bait, bet, bat,* and (for another reason) *bent* all tend to be spelled alike: BAT. In questioning children during various experiments, I have observed that they regard this homography, when it is pointed out to them, as perfectly acceptable and unremarkable. This attitude is just one illustration of a general property of children's early writing, namely that it seems to be created without much consideration of the needs of a reader. Another example is the way in which the phonetic spelling tends to conceal word relationships: the spellings THAQ for *thank you* or WRX for *works,* for instance, make it more difficult for the reader to recover the morphemes that make up these expressions. I have also known children to write messages to other children whom the writer knows cannot read. In all these ways, it seems that children's awareness of the function of their writing is at first limited to the writer's point of view. Their early writing need not be communicative in intent or achievement.

Finally, these phonetic judgments, like the segmentation tasks, are subject to a significant degree of individual variation. As Lila Gleitman emphasizes, some children have great difficulty in segmentation, while others (including those who make up their own spelling) have no trouble at all. This variation in development promises to make research on children's linguistic judgments somewhat messy, and underlines the importance of creating the best experimental designs for eliciting judgments. In this, we are not alone. Several areas of linguistic inquiry are coming to recognize individual and social variation, even in adult judgments of grammaticality. Furthermore, this messiness carries its own reward: it appears to be precisely those tasks which are subject to individual variation, such as learning to read and write, learning a second language, and learning to use language effectively in varied situations, which are of considerable practical importance. As our research takes account of individual differences, identifying the structure of linguistic judgments despite differences in rate of development, we may lay the foundation for better teaching and remediation.

*Other Kinds of Phonological Awareness*

Finally, it is tantalizing to observe that the relations of ambiguity and paraphrase which deserve to be studied on the syntactic level have counterparts on the phonological level. For example, non-significant phonetic contrasts are instances of the paraphrase relation at the phonological level. Thus cases of complementary distribution, such as the distribution of aspirated and unaspirated voiceless stops in English, or cases of free variation, such as the pronunciations [prIns] and [prInts] for *prince*(or *prints*) in English, are phonological paraphrases. More interestingly, puns constitute a special case of phonological near-ambiguity. They involve the replacement of an expected word with a substitute that meets two requirements: the substitute is somehow appropriate in meaning, preferably with an ironic twist, and it is phonologically similar, but not necessarily identical, to the expected word. A colleague of mine, for example, has hanging on his office wall an elaborate needlepoint production of the sentence, *A good pun deserves to be drawn and quoted*. The effectiveness of the pun depends on the degree of phonological similarity between *quoted* and *quartered*, of course. One wonders how children's awareness of such relationships develops and whether these developments are tied to their syntactic counterparts. With respect to phonological paraphrase, when do children become aware of individual differences in pronunciation? Is it the case, as it seems to be, that children's ability to appreciate puns and riddles develops at about age five or six, at roughly the same age as other, more traditionally recognized types of linguistic awareness?

Undoubtedly the greatest importance of studying phonological awareness, however, lies in its relationship to reading and writing, where an awareness of segments seems to be extremely useful if not absolutely essential for learning. It is much more difficult to show that linguistic awareness is really necessary to speaking and listening, unless one considers what is needed for rather high levels of effectiveness in spoken language. Not only are reading and writing activities for which a degree of linguistic awareness seems to be necessary, but they pose some profound problems. The choice of an orthography, for example, seems to depend in part on deciding just what constitutes *significant* variation in language. My research on children's spelling has been made more interesting by this type of question, and by the particular puzzles posed for the would-be writer by the phonology of English and its standard orthography.

The observation that phonological awareness is important for reading and writing ought to remind us to look beyond its practical utility to its broader intellectual and social importance. We must consider, as George Miller has urged us to do, the consequences of that degree of linguistic awareness (and self-awareness) required for writing. Miller relates the development of writing to the development of logic, for instance (1972, pp. 373-375). Contrary to the argument that I posed at the beginning of this paper, the study of linguistic awareness is not merely the analysis of a quaint but rather useless human performance which happens to be of service in refining psycholinguistic description but plays no observable role in ordinary language use and language development. On the contrary, our subject is a crucial part of that self-awareness and capacity for self-criticism which may underlie our sense of our own history and our own thinking.

Perhaps our awareness of some aspects of our language even forms part of the basis for our sense of our humanity and our worth. Consider again the triumphant enthusiasm with which a six-year old discovers his or her capacity to appreciate puns and riddles. It is true that the jokes are fun, but they may have a greater significance for the child's sense of himself or herself as a manipulator of language, for his or her recognition of the distance, and hence the room for play, between phonemic percept and semantic concept.

As the language-learning chimpanzees force us to examine more carefully just what it is that sets our linguistic abilities apart, one is tempted to inquire about the chimps' linguistic awareness: does Washoe ever pun by using one sign for another which resembles it physically but which differs in meaning? Does she experience pleasure from such activity? Linguistic awareness is not merely a source of data for structural description. The awareness itself is part of the human capacity for language, perhaps an important part, probably a distinctive part. Its form and development are worthy of study.

*FOOTNOTE*

1. I am grateful to Dennis Stampe for enlightenment on this point and other issues related to awareness.

*REFERENCES*

Bruce, D.J. The analysis of word sounds by young children. *British Journal of Educational Psychology*, 1964, 34, 158-170.

Campbell, H.W. Hierarchical ordering of phonetic features as a function of input modality. In: G.B. Flores d'Arcais & W.J.M. Levelt (eds.), *Advances in psycholinguistics*. Amsterdam: North-Holland, 1970.

Downing, J., & Oliver, P. The child's conception of a word. *Reading Research Quarterly*, 1973-1974, 9, 568-582.

Finnegan, D. *Phoneme sequencing knowledge of kindergarten and elementary school age children*. (Doctoral dissertation, University of Wisconsin, 1976). *Dissertation Abstracts International*, 1977, 37, 4922A. (University Microfilms No. 76-25, 558).

Fodor, J., Bever, T., & Garrett, M. *The psychology of language*. New York: McGraw-Hill, 1974.

Gleitman, L.R., Gleitman, H., & Shipley, E.F. The emergence of the child as grammarian. *Cognition*, 1972, 1, 137-163.

Graham, L.W., & House, A.S. Phonological opposition in children: A perceptual study. *Journal of the Acoustical Society of America*, 1971, 49, 559-566.

Hockett, C. *A course in modern linguistics*. New York: MacMillan, 1958.

Hayes, J.R., & Clark, H.H. Experiments on the segmentation of an artificial speech analogue. In: J.R. Hayes (ed.), *Cognition and the development of language*. New York: John Wiley, 1970.

Johns, J.L. A study on the child's conception of a spoken "word." In: W.O. Miller & G.H. McNinch (eds.), *Reflections and investigations on reading: Twenty-fifth yearbook of the National Reading Conference*. Clemson, South Carolina: The National Reading Conference, 1976.

Karpova, S.N. The preschooler's realization of the lexical structure of speech. In: F. Smith & G.A. Miller (eds.), *The genesis of language: a psycholinguistic approach*. Cambridge, Mass.: M.I.T. Press, 1966.

Liberman, I.Y., Shankweiler, D., Liberman, A.M., Fowler, C., & Fischer, F.W. Phonetic segmentation and recoding in the beginning reader. In: A.S. Reber & D. Scarborough (eds.), *Toward a psychology of reading*. Hillsdale, N.J.: Lawrence Erlbaum Associates, 1977.

McNinch, G. Experiments in phoneme shifting perceptions in pre-literate and literate samples. In: G.H. McNinch & W.D. Miller (eds.), *Reading: Convention and inquiry: Twenty-fourth yearbook of the National Reading Conference*. Clemson, South Carolina: The National Reading Conference, 1975.

Menyuk, P. Children's learning and reproduction of grammatical and non-grammatical phonological sequences. *Child Development*, 1968, 39, 849-859.

Messer, S. Implicit phonology in children. *Journal of Verbal Learning and Verbal Behavior*, 1967, 6, 609-613.

Miller, G.A. Reflections on the conference. In: J.F. Kavanagh & I.G. Mattingly (eds.), *Language by ear and by eye*. Cambridge, Mass.: M.I.T. Press, 1972.

Miller, G.A., & Nicely, P.E. An analysis of perceptual confusions among some English consonants. *Journal of the Acoustical Society of America*, 1955, 27, 338-352.

Read, C. Children's judgments of phonetic similarities in relation to English spelling. *Language Learning*, 1973, 23, 17-38.

Read, C. Children's categorization of speech sounds in English. *National Council of Teachers of English Research Report No. 17*, 1975.

Savin, H.B. What the child knows about speech when he starts to learn to read. In: J.F. Kavanagh & I.G. Mattingly (eds.), *Language by ear and by eye*. Cambridge, Mass.: M.I.T. Press, 1972.

Singh, S. Perceptual similarities and minimal phonemic differences. *Journal of Speech and Hearing Research*, 1971, 14, 113-124.

de Villiers, J.G., & de Villiers, P.A. Competence and performance in child language: Are children really competent to judge? *Journal of Child Language*, 1974, 1, 11-22.

Wickelgren, W.A. Distinctive features and errors in short-term memory for English vowels. *Journal of the Acoustical Society of America*, 1965, 38, 583-588.

Wickelgren, W.A. Distinctive features and errors in short-term memory for English consonants. *Journal of the Acoustical Society of America*, 1966, 39, 338-398.

Zhurova, L.Ye. The development of analysis of words into their sounds by preschool children. In: C.A. Ferguson & D.I. Slobin (eds.), *Studies of child language development*. New York: Holt, Rinehart and Winston, 1973.

# Aspects of Linguistic Awareness Related to Reading

Ingvar Lundberg

Institute of Psychology, University of Umeå, Umeå, Sweden

The author of the *Iliad* and the *Odyssey* describes an almost colorless world.
So Gladstone once proposed in a London pub that color vision had evolved in
mankind during historical time. I have been told that color blindness or
daltonism was not described in literature until the end of the seventeenth
century. Of course, the physical and physiological prerequisites for color
vision have not changed essentially in the course of history. But in one
sense Gladstone was probably right. There is a difference between sensation
and conscious awareness of or reflection on the phenomenon of color. Our
way of viewing color might well have changed since the days of Homer. An-
tique man probably saw the olive in the same manner as we do. For him,
however, there was no need to differentiate olive green from the green olive.
What use was there for having two words for what was one and the same thing
in inseparable union? Why have a separate name for the color? Did he even
notice it as color? To conceptually separate the color from the colored
object requires more abstraction than we usually presume. Surely this operation is much facilitated by practical work with colors. The techniques
of dyeing and painting have probably been of importance in the evolution of
color terms. And these techniques have assuredly been developed during historical time.

In the field of language we have a similar situation. Piaget (1926)
told us that the illiterate people of Golah in Liberia were ignorant of the
fact that their language consisted of words; the real unit for them was the
phrase or the sentence. I think it is relevant to consider the young child's

capacity against mankind's historical perspective. Human beings have been largely illiterate throughout history. Though logographic systems appeared far earlier and have been constructed independently several times in different varieties (Gelb, 1963), it was the alphabetic system that brought about a psychological revolution. This came into being only around 1500 B.C., which means that it took man more than half a million years to invent it. And it is indeed a remarkable construction with almost perfect economy of means. Now, this elegant information economy, as we will see, was bought for a price which was high for many people. For the first time in history, planned and systematic education became a compelling need. For the basis of the alphabetic system, the phoneme, is perceptually and conceptually not easily available.

To a preschool child it is in no way apparent that language consists of words, that words vary in length, that words are built up from parts, and the like. "Synthesis of an utterance is one thing; the awareness of the process of synthesis quite another" (Mattingly, 1972, p. 140). In normal and natural communication we encounter little need to focus upon anything but the meaning of utterances. The language forms are themselves transparent. We hear through them to the meaning intended. Polanyi (1963) notes that "if you shift your attention from the meaning of the symbol to the symbol as an object viewed in itself, you destroy its meaning.... Symbols can serve as instruments for meaning only by being known subsidiarily while fixing our focal attention on their meaning" (p. 30). Although literate adults are capable of dual knowledge of the "figure" and "ground" aspects of language, we should not expect young children to be conscious of their own language functions in any formal sense. Learning to read, however, should bring about a major change in the child's metalinguistic knowledge.

*THE PRESCHOOL CHILD'S CONCEPTION OF WORDS*

That the child has difficulty with words when they are abstracted from the flow of speech was pointed out by Vygotsky (1962). He observed that young children treat the names of objects as if they were intrinsic properties. "When asked whether one could interchange the names of objects, for instance call a cow 'ink' and ink 'cow,' the child will answer no, 'because ink is used for writing and the cow gives milk.' An exchange of names would mean an exchange of characteristic features, so inseparable is the connection between them in the child's mind" (Vygotsky, 1962, p. 129). Ianco-Worrall

(1972) observed that a group of bilingual children separated sound from meaning in words earlier than a matched group of monolingual children. To study attention to meaning or sound, she used a forced-choice task in which similarity between words could be interpreted on the basis of shared meaning or shared acoustic properties. She also applied the Vygotskyan interview technique to study the arbitrary nature of the name-object relationship. It was found that this notion developed later than the ability to separate the qualities of objects from their names.

The poor word concept of the preschool child was also noted by Papandropoulou and Sinclair (1974). When asked to say a long word young children (aged 4;6 to 5;6) answered with names for long and big objects or expressions for actions that take a long time and vice versa for short words, while older children (aged 7 to 8) produced phonologically long and short words regardless of semantic content. The child's sensitivity to word length was also studied by Rozin, Bressman, and Taft (1974). They designed a task where the child was shown a pair of written words, one short and one long. Both words were given orally as well. Though unable to read, the child was asked to pick one of the written words when it was spoken by the experimenter. The basis for a correct solution to this problem must be that the child (who cannot read) discriminates acoustical word length and maps this onto the length of the letter display. Consideration of articulatory complexity might also be involved. To the literate adult the sound-orthography relation is so obvious that it is difficult to appreciate the problems a child might have in attending to the temporal and phonological characteristics of spoken words, and in relating them to graphic length and complexity of written words.

Lundberg and Tornéus (1977), using a similar choice task, studied Swedish children four to seven years old. In their study, in addition to referent size, word length was varied in three ways--by number of letters (and phonemes), by length of vowels, and using compound vs. simple nouns. Physical length differences were balanced across conditions. Semantic congruence (the relation of word to referent size) and noun compounding both had main effects. The probability of a correct choice was greater when the graphically long word denoted a large object and the short word denoted a small object (e.g., *ambulance - arm*). However, responses of this type indicate that the child simply stays in the semantic mode and doesn't shift his attention to word length proper. A genuine understanding of the sound-orthography relation can only be revealed in the reverse condition. Here the short word denotes a large object and the long word a small object (e.g., *tree - tele-phone*). If a subject now makes a correct choice, he is (with due account for guessing) disregarding the semantics and attending to length of the words per se. Children who show this kind of resistance to counter-suggestion were found only in the older age groups (6 and 7 years). In the youngest group (aged 4 years) irrelevant and non-linguistic solutions predominated, and semantic content was also a powerful guide for five-year olds, although in both groups compound words were judged to be longer than simple ones. But only the oldest children showed proper understanding of the basic relationship between spoken and written words.

A binary choice method was also used by Downing and Oliver (1974). The children played a "yes-no" game in which various types of auditory stimuli were presented. They were requested to respond "yes" if they thought they heard a word and "no" if they did not. The results suggest that young children (up to the age of 8 years) do not have an adequate concept of what constitutes a spoken word, a finding which should be recognized by reading teachers. Most children, according to Downing and Oliver, enter school in a state of "cognitive confusion" over the components of language.

Ehri (1975), Holden and McGinitie (1972), Huttenlocher (1964), and Karpova (1966) have all found preschool children unable to segment meaningful sentences into their component words. They are particularly likely to ignore function words. Typically (e.g., Holden & McGinitie, 1972), children were asked to segment phrases and sentences into words by tapping for each word spoken. However, this technique may be unduly complex, since it requires the child to perform the simultaneous tasks of repeating utterances and identifying their word boundaries. There are also reasons to believe that the rhythmic pattern of sentences governs the marking behavior of many children. Normally, content words are given more stress; it is possible that this causes them to mark contentives and ignore functors.

As we have seen, the child's conception of words involves several aspects, and various techniques have been used to study them, including clinical interviews, choice paradigms, and sentence segmentation tasks. As long as this problem is treated in isolation and theoretical relationships to other aspects of cognitive and linguistic development are left vague, however, the situation will remain inconclusive and perhaps confusing. An adequate word concept is just one dimension of linguistic awareness. Let us explore some further aspects.

*THE ATTENTION SHIFT*

The central aspect of linguistic awareness is an *attention shift* from content to form, the ability to make language forms opaque. This seems to be a special kind of language performance, less easily and less universally acquired than normal speaking and listening. Certainly very young children *playfully* manipulate speech sounds apart from their meanings. Cazden (1975) noted that "children may shift more easily than adults between using language forms transparently in interpersonal communication, and treating them as opaque objects in play." But this kind of playful activity may be essentially different from cognitively more demanding skills involved in conscious metalinguistic reflection, analysis, evaluation, or synthesis.

Sinclair de Zwart (1971) suggests that changes in *thinking* around six or seven lead to new forms of verbal behavior. A general capacity for detachment paves the way for metalinguistic reflection, as well as the ability to find different verbal formulations describing the same event. Now children can conserve semantic content while changing the form of an utterance. Unfortunately, hard data are missing which support this view. The important point, however, is that linguistic awareness be studied in relation to cognitive development in general.

Taking Piaget's decentering theory as a point of departure we are currently exploring the relationship between linguistic awareness and perceptual *decentration*. The latter is being measured by a picture integration test used by Elkind, Koegler and Go (1964) where parts and wholes can be perceived in drawings wherein each has independent meaning, e.g., a man composed of pieces of fruit. We have also constructed a target identification test where the child has to point out a formally defined target in a meaningful pictorial context. The target is a geometric shape with no intrinsic meaning. For example, it may be part of a meaningful object seen in depth, such as a roof (see Figure 1), or the empty space between objects in different depth positions. The parallel to typical tasks used to assess linguistic awareness is almost complete. The same kind of attention shift is demanded between meaningful content and formal properties. However, in a recent study of two hundred preschool children, the statistical correlation with a phoneme analysis task was not higher than +.30. Studies of relationships to other kinds of task are now underway.

Before closing this section we should note that the attention shift may occur in steps, or at different levels going from opaque meaning to transpar-

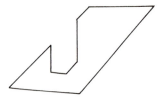

Figure 1.

ent form. McNeill (1974) formulated the following principle: "whatever is last (or largest) to be processed in speech perception is opaque, and is in the focus of attention, and everything preliminary (or smaller) is transparent...rather than a dichotomy between opaque meaning and transparent syntax or phonology, there is a series of opaque-transparent oppositions, depending on how far linguistic processing has advanced before it stopped" (p. 224). With this principle in mind we can now consider methods for assessing linguistic awareness.

## TASK DIMENSIONS

A multitude of methods have been used to assess aspects of linguistic awareness. By considering the following three dimensions a reasonable taxonomy of tasks may be obtained:

(1) Cognitive operation involved (e.g., synthesis or analysis)

(2) Size of units (e.g., phrases, words, syllables, phonemes)

(3) Amount of mental activity or complexity of processing required by the task (e.g., compare the rather passive resonance in recognizing rhyme with the mental effort involved in phoneme segmentation).

### Synthesis Tasks

The subject is presented with parts or fragments of a linguistic unit and is requested to integrate them back into the original "gestalt." In tests of auditory or grammatical *closure* (Kirk, McCarthy, & Kirk, 1968) the gestalts used have comparatively small defects and the closure mechanism normally operates with minimal cognitive demands. But in a typical synthesis task the breakdown in the gestalt is more pronounced and integration requires a corresponding analysis of the total unit in the subject's lexicon, the result of which is matched against the presented fragments.

*Logographic phrase integration.* In a study by Farnham-Diggory (1967), four- and five-year olds learned a set of eight logographs--strongly schematized figures representing words such as *jump* and *around*. In a test phase the child was presented with a logograph sentence, e.g., *Jump around teacher*. The child read word by word with no difficulty but, when the experimenter said "do it," looked confused. At best he could jump up and down, make a circular hand movement, and then point to the teacher. Why cannot the child integrate the units? A similar task is quite easily managed by the chimpanzee Sarah (Premack, 1971). What is the reason for this unexpected difference between child and animal? Ferguson (1975) has shown that reading achievement is highly predictable from performance of logographic tasks of the kind described here. Perhaps the chimpanzee has the advantage of not being bound by a natural, articulated language. She has no semantic prison to escape from. Her only linguistic universe consists of plastic symbols. But the child's starting point is a natural phrase, a gestalt which is broken down by the segmentation. A prerequisite for re-integrating or cognitively synthesizing the separate elements may be an insight, an awareness of the fact that sentences and phrases in the language consist of words.

*Word synthesis or sound blending.* Auditory blending occurs when a word is produced by synthesizing its component elements (syllables or phonemes) heard separately. There is a large area of testing devoted to examining the skill of auditory blending (Kirk, McCarthy, & Kirk, 1968; Roswell & Chall, 1963). The relationship between auditory blending and success in reading is also well documented (Chall, Roswell, & Blumenthal, 1963; Flynn & Byrne, 1970; Hardy, Stennett, & Smythe, 1973; Reynolds, 1963). The early development of auditory blending, however, has not been extensively investigated. Goldstein (1976) and Helfgott (1976) are two recent exceptions.

Coleman (1970) reported that it was easier for subjects to blend together VC (vowel-consonant) than CV syllables and to blend continuant rather than stop consonants. However, in a recent study by Haddock (1976) these results were not confirmed. To clarify this issue more research is needed. Helfgott (1976) compared CV-C and C-VC tasks. Greater ease of blending found for CV-C items was explained with reference to the structure of the syllable. In spoken language the consonant-vowel syllable seems basic (Bondarko, 1969).

To be able to blend a word from its phonetic components, the child must understand the relationship between phonemes in isolation and phonemes in words. Isolating the phoneme, however, is a serious conceptual problem. A word is built up from phonemic segments. But the problem is that these segments are abstract. Acoustically there is just one gestalt. When we hear a word, we never hear in fact the sequence of segments of which the word consists. What we perceive is rather the result of various transformations of the sequence (e.g., Chomsky & Halle, 1968). Thus, the form of a word has two sides. Its inner side is an abstract representation, a sequence of

phonemes. Its outer side is a phonetic representation derived from the abstract form by phonological rules. By co-articulation these phonetic segments themselves are coded into sound envelopes of approximately syllable
size in which phonemes are transmitted in parallel (Liberman, 1970).

It is reasonable to assume that the inner side is not easily accessible
to introspection by the child. In fact, it took mankind hundreds of thou-
sands of years to codify it. This problem is often overlooked in beginning
reading instruction when a teacher naively assumes that a proper synthesis
will be induced simply by rapidly blending the elements. But the understanding needed cannot be achieved until the child becomes aware of the abstract
phonemic structure. How this awareness is developed remains largely unclear.

*Analysis*

Auditory segmentation involves the ability to segment utterances of
various types and lengths into subunits of various types and sizes, e.g.,
sentences into words, words into syllables, and syllables into phonemes. The
segmentation of phrases into words has already been discussed. Though even
infants can functionally distinguish different speech sounds (Eimas, Siqueland, Jusczyk, & Vigorito, 1971), using this capacity to consciously analyse
words into sounds is a much later achievement. A number of recent investigators have taken the view that phonemic segmentation is an essential skill
in beginning reading that "reading-disabled" children have failed to acquire
(Calfee, Lindamood, & Lindamood, 1973; Gleitman & Rozin, 1973; Helfgott, 1976;
Liberman, 1973; Mattingly, 1972; Rosner, 1974). Several different types of
tasks have been used to study syllabic or phonemic segmentation.

*Partial segmentation.* Identifying initial or final consonants (Zhurova,
1973) and rhyming are examples of tasks requiring only partial segmentation.
Johnson and Myklebust (1967) report clinical observations of dyslectic children who even at the age of 12 are insensitive to rhyme, although they seem to
have normal hearing and speech. To appreciate a rhyme does not seem to require a very analytic attitude. Such thinking is more involved in initial
and final phoneme segmentation, and cognitive load is still more pronounced
in an elision procedure where the child is asked to delete part of a word
and say what remains, for example, to say *desk* without /s/. Such a procedure
was used by Bruce (1964), who concluded that children below the mental age
of seven could not perform phonemic analysis of spoken words. However, the

task seems unduly complex for justifying this conclusion. Rosner (1974) gave
a group of non-reading grade one children auditory training daily with the
goal of teaching them to add, omit, substitute, or rearrange the phonemic
elements of spoken words. Compared to a control group given no similar training the experimental group was superior in auditory analysis, as well as in
word recognition. In preliminary studies in Umeå we have in addition to the
elision procedure used a corresponding task where the child is given the
part remaining from, e.g., a word and asked to say what element has been
deleted. For example, "take the word *post*. Now I say /pot/. What was lost?"
The position as well as size and type of the element was varied. The two
elision variants were compared with a corresponding closure task.

The complexity of analysis in such tasks seems to be determined by the
following factors: (1) the size of the starting unit (e.g., phrases, multi-
syllabic words, monosyllabic words); (2) the size of the element under analysis; (3) the context in which the elements are embedded (e.g., consonants
in clusters vs. in simple vowel-consonant combinations); and (4) the relative
position of the element within the starting unit. Thus, the anatomy of words
so intensively studied within the tachistoscopic tradition is also displayed
in this context. The relevant underlying processes may concern memory and
attention as well as mechanisms of language perception.

Most *secret languages* require some partial phoneme segmentation. Day
(1973) introduced the secret language context for studying psycholinguistic
phenomena and looked at individual differences among adults. Savin (1972)
noted that children with reading disability were unable to learn Pig Latin,
although they were highly motivated. Besides segmentation, this task seems
to require mental operations such as displacement and addition.

*Complete segmentation.* Full segmentation was studied by Liberman (1973),
who had subjects tap on a table to indicate one, two or three phonemes in a
word. In a procedure used by Elkonin (1973) and later adopted by Helfgott
(1976) subjects are presented with a picture and the spoken name of the depicted object. The task is to say the phonemes and synchronously move a
counter into the squares of a visual model of the word. The results have
shown that test items are more readily segmented into syllables than into
phonemes. There is also a clear developmental trend. In the study by Liberman (1973), phoneme segmentation did not appear until the age of five, and
then was demonstrated in only 17 percent of cases. Among six-year olds 70

percent succeeded in phoneme segmentation, while 90 percent were successful
in the syllable task. It is still unclear whether increasing ability to segment phonetically is a result of reading instruction or whether it is a more
general manifestation of cognitive growth. The latter alternative receives
some support from our investigation in Sweden, where children do not start
school until the age of seven.

*THE RELATIONSHIP TO READING DISABILITY*

To account for poor reading performance of lower class children, Savin
(1972) points to different metalinguistic experiences. Whereas phonemic
analysis is something that middle-class children are likely to be familiar
with, probably having engaged in activities such as rhyming and other language games, inner-city children may often come to school prepared to use
only syllabic analysis of the speech they hear. We think, however, that the
relationship between reading disability and metalinguistic competence is a
bit more complicated.

The experimental and educational literature is crowded with studies
where normal readers have been compared with dyslectic children in a great
number of respects. Sometimes significant differences have been observed
and sometimes not. Contradictory results are not uncommon. Treating heterogenous groups as if they were homogeneous inevitably yields such outcomes.
Certainly there is a high risk that straightforward and simple relationships
will be obscured. The heterogeneity of the reading-disabled has recently
been recognized by Gjessing (1977), a Norwegian researcher. On the basis of
thorough and long-term clinical observations of a great number of cases he
describes six diagnostic subgroups of dyslexia, of which the most important
are auditory and visual dyslexia. The typical child with auditory dyslexia
displays serious problems with phonemic analysis and synthesis, which are
manifested in reading and writing. The child with visual dyslexia suffers
primarily with visual memory problems, manifested by a different pattern of
results. We are now involved in a cross-national project within Scandinavia
on learning disabilities. One of the many objectives of the project is to
further test and elaborate Gjessing's theory of dyslexia. In Umeå we are
measuring various aspects of the metalinguistic competence of a group of preschool children. We then follow them up during the first three school years
and observe how different aspects of metalinguistic skills are related to
different disability patterns. Both the practical and theoretical implica-

tions of such data might be of value.

*CONCLUDING COMMENTS*

The present review has been limited primarily to phonological aspects of metalinguistic competence. Here we have seen that a multitude of methods for assessment have been introduced in the last decade. However, the attempt to create a taxonomy does not dispel the impression of a rather ill-defined field. A number of significant issues remain unclear. For example, what is the relationship between synthesis and analysis? Is there a reciprocal relationship between learning to read and the developmental changes in metalinguistic competence? What are the central dimensions in linguistic awareness, and are there qualitatively different stages in its development? How easily can the development of linguistic awareness be modified by stimulating experiences, and which factors are most pertinent in such stimulation? How is linguistic awareness related to social and emotional development, e.g., self-concept? And, most generally, how is linguistic awareness related to cognitive development in general?

*REFERENCES*

Bondarko, L.V. The syllable structure of speech and distinctive features of phonemes. *Phonetica*, 1969, 20, 1-40.

Bruce, L.J. The analysis of word sounds by young children. *British Journal of Educational Psychology*, 1964, 34, 158-170.

Calfee, R.C., Lindamood, P., & Lindamood, C. Acoustic-phonetic skills and reading--kindergarten through twelfth grade. *Journal of Educational Psychology*, 1973, 64, 293-298.

Cazden, C.B. Play with language and meta-linguistic awareness: One dimension of language experience. In: J.S. Bruner, A. Jolly, & K. Sylva (eds.), *Play: Its role in evolution and development*. London: Penguin, 1975.

Chall, J., Roswell, F.G., & Blumenthal, S.H. Auditory blending ability: A factor in success in beginning reading. *Reading Teacher*, 1963, 17, 113-118.

Chomsky, N., & Halle, M. *The sound pattern of English*. New York: Harper & Row, 1968.

Coleman, E.B. Collecting a data base for a reading technology. *Educational Psychology Monographs*, 1970, 61, 1-23.

Day, R.S. On learning "Secret languages." Haskins Laboratories, *Status Report on Speech Research*, SR-34, 1973.

Downing, J., & Oliver, P. The child's conception of 'a word.' *Reading Research Quarterly*, 1973, 1974, 9, 568-582.

Ehri, L.C. Word consciousness in readers and prereaders. *Journal of Educational Psychology*, 1975, 67, 204-212.

Eimas, P.D., Siqueland, E.R., Jusczyk, P., & Vigorito, J. Speech perception in infants. *Science*, 1971, 171, 303-306.

Elkind, D., Koegler, R.R., & Go, E. Studies in perceptual development: II. Part-whole perception. *Child Development*, 1964, 35, 81-90.

Elkonin, D.B. U.S.S.R. In: J. Downing (ed.), *Comparative reading*. New York: MacMillan, 1973.

Farnham-Diggory, S. Symbol and synthesis in experimental reading. *Child Development*, 1967, 38, 221-231.

Ferguson, N. Pictographs and prereading skills. *Child Development*, 1975, 46, 786-789.

Flynn, P.T., & Byrne, M.C. Relationship between reading and selected auditory abilities of third-grade children. *Journal of Speech and Hearing Research*, 1970, 13, 731-740.

Gelb, I.J. *A study of writing*. Chicago: Chicago University Press, 1963.

Gjessing, H.J. *Dyslexi*. Oslo: Universitetsforlaget, 1977.

Gleitman, L.R., & Rozin, P. Teaching reading by use of a syllabary. *Reading Research Quarterly*, 1973, 8, 447-483.

Goldstein, D.M. Cognitive-linguistic functioning and learning to read in preschoolers. *Journal of Educational Psychology*, 1976, 68, 680-688.

Haddock, M. Effects of an auditory and an auditory-visual method of blending instructions on the ability of prereaders to decode synthetic words. *Journal of Educational Psychology*, 1976, 68, 825-831.

Hardy, M., Stennett, R.G., & Smythe, P.C. Auditory segmentation and auditory blending in relation to beginning reading. *The Alberta Journal of Educational Research*, 1973, 19, 144-158.

Helfgott, J.A. Phonemic segmentation and blending skills of kindergarten children. Implications for beginning reading acquisition. *Contemporary Educational Psychology*, 1976, 1, 157-169.

Holden, M.H., & McGinitie, W.H. Children's conceptions of word boundaries in speech and print. *Journal of Educational Psychology*, 1972, 63, 551-557.

Huttenlocher, J. Children's language: Word-phrase relationship. *Science*, 1964, 143, 264-265.

Ianco-Worrall, A.D. Bilingualism and cognitive development. *Child Development*, 1972, 43, 1390-1400.

Johnson, D., & Myklebust, H. *Learning disabilities: Educational principles and practices*. New York: Grune & Stratton, 1967.

Karpova, S.N. Abstracted by D.I. Slobin in: F. Smith & G.A. Miller (eds.), *The genesis of language*. Cambridge, Mass.: M.I.T. Press, 1966.

Kirk, S.A., McCarthy, J., & Kirk, W. *Examiner's manual: Illinois Test of Psycholinguistic Abilities* (rev. ed.). Urbana, Ill.: University of Illinois Press, 1968.

Liberman, A.M. The grammars of speech and language. *Cognitive Psychology*, 1970, 1, 301-323.

Liberman, I.Y. Segmentation of the spoken word and reading acquisition. *Bulletin of the Orton Society*, 1973, 23, 65-77.

Lundberg, I., & Tornéus, M. Nonreaders' awareness of the basic relationship between spoken and written words. (submitted for publication, 1977).

Mattingly, I.G. Reading: The linguistic process and linguistic awareness. In: J.F. Kavanagh & I.G. Mattingly (eds.), *Language by ear and by eye: The relationships between speech and reading*. Cambridge, Mass.: M.I.T. Press, 1972.

McNeill, D. How to resolve two paradoxes and escape a dilemma: Comments on Dr. Cazden's paper. In: K.S. Connolly & J.S. Bruner (eds.), *The growth of competence*. New York: Academic Press, 1974.

Papandropoulou, I., & Sinclair, H. What is a word? Experimental study of children's ideas on grammar. *Human Development*, 1974, 17, 241-258.

Piaget, J. *The language and thought of the child*. London: Routledge and Kegan Paul, 1926.

Polanyi, M. *The study of man*. Chicago: Chicago University Press, 1963.

Premack, D. Language in chimpanzee? *Science*, 1971, 172, 808-822.

Reynolds, M.C. A study of the relationships between auditory characteristics and specific silent reading abilities. *Journal of Educational Research*, 1963, 46, 439-449.

Rosner, J. Auditory analysis training with prereaders. *The Reading Teacher*, 1974, 27, 379-384.

Roswell, F.G., & Chall, J. *Auditory blending test*. New York: Essay Press, 1963.

Rozin, P., Bressman, B., & Taft, M. Do children understand the basic relationship between speech and writing? The mow-motorcycle test. *Journal of Reading Behavior*, 1974, 6, 327-334.

Savin, H.B. What the child knows about speech when he starts to learn to read. In: J.F. Kavanagh & I.G. Mattingly (eds.), *Language by ear and by eye*. Cambridge, Mass.: M.I.T. Press, 1972.

Sinclair de Zwart, H. Piaget's theory and language acquisition. In: *Piagetian cognitive developmental research and mathematical education*. Washington: N.C.T.M., 1971.

Vygotsky, L.S. *Thought and language*. Cambridge, Mass.: M.I.T. Press, 1962.

Zhurova, L.Ye. The development of analysis of words into their sounds by preschool children. In: C.A. Ferguson & D.I. Slobin (eds.), *Studies of child language development*. New York: Holt, 1973.

# What Did the Brain Say to the Mind? A Study of the Detection and Report of Ambiguity by Young Children

Kathy Hirsh-Pasek, Lila R. Gleitman, and Henry Gleitman

Graduate School of Education, University of Pennsylvania
Philadelphia, PA 19104, USA

Verbal humor is very much a part of everyday life, so it is not surprising that even young children laugh at riddles and jokes. They also invent their own, but these are perplexing. Here is an example from a five-year old of our acquaintance:

      Child: What has a trunk and four wheels?
      Us   : I don't know. What has a trunk and four wheels?
      Child: A car! (hilarious laughter)

Despite manifest facility with the riddle format, this child apparently is unaware that riddles, at least *good* riddles, turn on linguistic ambiguity. At the same time, it is easy to show that children of this age are in perceptual and productive control of two senses for a single word (e.g., the two meanings of *bark* or *club*) or constructions (e.g., the two meanings of *flying planes*). This is one of many instances where children display competence in speech and understanding, but fail when the task is to see through to the linguistic event itself, manipulating it in the service of providing a judgment. Briefly, while the youngster is sensitive to potential alternate interpretations of speech signals, he cannot answer to the fact that some single speech event can be interpreted in two ways. He cannot give judgments concerning ambiguity.

      We have been studying such disparities between language use (linguistic

skill, or skill *at* language) and language judgments (metalinguistic skill,
or skill *about* language) for some years (for a summary account, see Gleitman
& Gleitman, in press).  For a variety of language issues, judgmental performance does lag behind speech and comprehension in developmental appearance.
Moreover, late appearance of such skills in the child is mirrored by the
degree of difficulty and extent of individual variation among adults.  Finally,
the difficulty of judgment giving is a function of the language level (e.g.,
syntax or semantics) on which the individual is being asked to report.  In
this paper, we try to describe an aspect of children's behavior in explaining
verbal jokes in terms of these prior findings on metalinguistic capacity.
In the second section we compare language perception and judgments, and review prior work.  The third section describes a study of children's appreciation of ambiguity in jokes.  The last section summarizes these outcomes in
terms of the postulated metalinguistic capacity.

## THE STRUCTURE OF LANGUAGE JUDGMENTS

The human ability to give judgments about well-formedness and sentential
relations has been studied intensively over the past twenty years by linguists in the tradition of generative grammar.  By and large, the data base
for grammar construction has been speakers' reports on three language dimensions: *classification* (judgments of grammaticality and internal structure of
sentences), *paraphrase* (judgments that two sentences code the same concept),
and *ambiguity* (judgments that a single sentence has two readings; i.e.,
codes more than one concept).  The achievements in language description,
working from these speaker reports, have been enormous, including the Standard
Theory of generative syntax (Chomsky, 1965) and a variety of revisions and
expansions that have followed.

However, a number of methodological and substantive attacks have been
mounted against the psychological relevance of this kind of work.  First,
there is the problem that the judgments on which the theory construction is
based are shaky and are not generally elicitable from all the individuals
whom we would want to characterize as, say, speakers of English.  Second,
and probably more serious, the descriptions of language built on the judgmental base do not accord in a simple way with the facts about human speech
and comprehension performance in the psychological laboratory (see, for
discussion, Fodor, Bever, & Garrett, 1974; Wanner, 1977).  On both grounds,
some theorists have argued that language judgments are epiphenomena that cannot

serve as the basis for a psychology of language. In an extreme view, some argue that such judgments, and hence the grammars constructed to describe them, have little or no interest for they are not psychologically "real." In a different move, some linguists have tried to broaden the data base for linguistic descriptions; that is, to write grammars responsive to facts about sentence perception as well as facts about judgments (e.g., Bresnan, in press).

We take the view that judgments of grammaticality, ambiguity, paraphrase, and the like, must be accounted for in a psychologically valid description of human language. Human consistency and scope in such tasks is enormous (for some experimental evidence, see Gleitman & Gleitman, 1970). Moreover, the use of judgmental data is commonly accepted within psychology, and has led to some of the most coherent and general theoretical achievements, most notably in the study of visual perception.

However, it is not necessarily useful to suppose that the psychological sources of language judgments and of speech performance are the same (for discussion, see Chomsky, 1965; Pylyshyn, 1973). Clearly, acts of speech and comprehension are in part manifestations of linguistic skills. We take judgments about language to be manifestions of an executive, or metalinguistic, skill that has psychological interest in its own right. The metalinguistic capacity show more individual and population difference than the linguistic capacity; it appears relatively late in development; and it is sensitive to linguistic levels. Specifically, the more "surface" aspects of language are more difficult to access for the sake of giving judgments than are the "deeper" or more meaningful aspects. This distinction in performance may reflect differences in decay rates for less and more highly processed linguistic material. Following, we review evidence on judgments for the linguistically central tasks of classification, paraphrase, and ambiguity.

*CLASSIFICATION*

George Miller and his associates (Miller, 1962; Miller & Isard, 1963; Marks & Miller, 1964) established the psychological relevance of a dimension related to grammaticality in a series of experiments on language perception and recall. They showed that such well-formed sentences as *Furry wildcats fight furious battles* are easier than their ill-formed relatives (e.g., *Furry fight furious wildcats battles*) to memorize, detect under degraded stimulus conditions, etc. Hence, there is a demonstrable effect in sentence perception of the grammatical classification of word-strings.

However, results are less than categorical if we query subjects directly about grammatical classification. Maclay and Sleator (1960) asked undergraduate subjects to classify sentences as to their grammaticality. Sensibly enough, 18 of their 21 subjects classified the monstrosity *Label break to be calmed about and* as ungrammatical. But three subjects classified it as grammatical. Yet surely these three, like the other subjects, would not utter such word-strings nor detect them easily in noise.

## Judgments of Grammaticality from Young Children

Shatz (1972) and Gleitman, Gleitman, and Shipley (1972) asked children to comment about anomalous sentences. The instructions were simply: "Tell me if these sentences are good or if they are silly." They found that children of five years of age were able to recognize and report on implausibilities and meaning anomalies in sentences. For instance, the sentence *The color green frightens George* was rejected on grounds that "Greens don't have faces of paint" or "Boys are used to green." But violations of syntactic form that scarcely affected meaningfulness usually went unnoticed by these kindergartners. For instance, in response to *John and Bill is a brother*, a child responded "Sure, they could be brothers, they're brothers." Notice that this child has demonstrated his productive control of verbal concord. In his own rendition, subject and verb agree in number. Yet he manifests no awareness that the stimulus sentence lacked this obligatory syntactic property.

On the contrary, seven-year old subjects in this same experimental context usually accepted semantically odd or implausible sentences as "good" and not "silly." For example, a subject responded to *The color green frightens George* by saying "Doesn't frighten me, but it sounds OK." Acknowledging the semantic implausibility, the subject has gone on to consider the sheer "sentencehood" of the stimulus. Similarly, in response to *Claire and Eleanor is a sister*, a seven-year old comments "You can't use *is* there: Claire and Eleanor *are* sisters." The paraphrase reflects the appreciation of meaning, but an explicit syntactic judgment is rendered as well. Sometimes these seven-year olds were capable of making rather nice partitionings between semantic and syntactic facts. In response to *I saw the queen and you saw one* (the indefinite pronoun is "wrong" if it is an anaphor of *queen*), a subject responded "Yeah, you saw 'a One,' whatever 'a One' is." (Since coreferentiality is denied, the sentence becomes grammatical.)

Overall these findings cohere on the view that younger children are able to spot and comment on meaning anomalies but not structural anomalies, while older children can give judgments on both meaning and form.

*Evidence from Reading Acquisition*

Many findings in the literature of reading acquisition support the claim that classificatory judgments appear earlier for deeper, more meaningful, or more global properties of language than for surface structural properties. Children of five can be taught the difference between the concepts "word" and "sentence" with little difficulty, but it is hard for them to distinguish, on the basis of similar teaching techniques, among the concepts "word," "syllable," and "sound," (Downing & Oliver, 1973-4). Children of five and six have mild difficulty segmenting speech into words (Holden & MacGinitie, 1972), but they have far greater difficulty in segmenting words into syllables (Rosner, 1974; Liberman, Shankweiler, Fischer, & Carter, 1974) and the greatest difficulty of all in segmenting words or syllables into phonemes (El'konin, 1973; Rosner & Simon, 1971; Rozin & Gleitman, 1977). In sum, the lower the level of linguistic representation called for in a judgmental task, the more difficult the task for young children. Perhaps the most dramatic evidence that this failure is not a failure of language perception comes from studies of phoneme discrimination, compared with phoneme classification, in young children.

Eimas, Siqueland, Jusczyk, and Vigorito (1971), using a dishabituation paradigm, showed that even four-week old infants will discriminate phonological properties of speech sounds relevant to language; they can discriminate, e.g., between *ba* and *pa*. While humans are not the sole possessors of such discriminations (Kuhl & Miller, 1975) and while humans can also discriminate categorically among acoustic stimuli not relevant to speech (Cutting & Rosner, 1974), the findings of Eimas et al. speak to the fact that the acoustic discriminative apparatus on which language learning is ultimately based is in place approximately from birth. Yet a well-known "reading readiness test" called the Auditory Discrimination Test (Wepman, 1958) is based on the fact that some kindergartners cannot correctly say "same" or "different" in response to pairs of words that differ in one phonological segment (e.g., *bat, pat*) or are identical (e.g., *bat, bat*).

From the demonstration of Eimas et al., we know that these children

can *hear* the differences in such stimuli. They can even correctly repeat
the stimulus items which they could not judge on the test (Blank, 1968).
Furthermore, the failing five-year olds do have the ability to give judgments: they can correctly say "same" and "different" in response to written
stimuli (e.g., they can discriminate between the visual displays BAT and
PAT, and judge them to be different, even though they cannot read them; Smith,
1973). Evidently the child who fails the Wepman test is very circumscribed
in his deficits. His weakness appears only when he is asked to give a judgment about the sound properties of linguistic stimuli. Yet the Wepman test
is a fairly accurate predictor of early reading success, suggesting that the
judgmental faculty is implicated in learning to read.

Rozin and Gleitman (1977) have argued that the requirement for conscious
awareness of surface language units is a major cognitive barrier to reading
acquisition. Speech perception and language learning take place without
awareness, in terms of an evolutionarily old and highly evolved mental cir-
cuitry. But for alphabetic reading, in particular, the learner is asked to
realize quite consciously that *pat* is decomposable into "p," "a," and "t,"
that *pat* starts with what *tap* ends with, and the like. The requirement for
conscious recognition predicts that learning to read should be more difficult
and more variable than learning to talk, as it is. Adequate speech is acquired, though slowly, even by retardates (Lenneberg, 1967; Lackner, 1976;
Morehead & Ingram, 1976). Despite differences in the language being learned
and in cultural ambiance, normal children seem to pass through similar sequences of developmental accomplishments during the same narrow time frame
(Brown, 1973; Slobin, 1973, 1975). Spoken language seems to emerge more or
less equivalently under a variety of content and presentation conditions;
that is, many aspects of language learning are insensitive to differences in
caretaker speech styles (Newport, Gleitman, & Gleitman, 1977). Successful
language-like means are achieved even by deaf children radically deprived of
linguistic input (Feldman, Goldin-Meadow, & Gleitman, 1978). In contrast,
the success and scope of reading acquisition vary as a function of intelligence (Singer, 1974), motivational and cultural factors (Downing, 1973) and
differences in the script that is to be learned (Rozin, Poritsky, & Sotsky,
1971).

A number of studies show, as we would predict, that it is the surface
properties of language to be accessed (i.e., the phonological substrate)
that causes the trouble in learning to read, not the meaningful content.

Firth (1972) showed that groups of third graders matched for IQ, but differing in reading skills according to the estimates of their teachers, performed identically on such semantic tasks as guessing plausible completions of incomplete orally presented sentences; but the ability to provide consensual pronunciations for written nonsense words (such as *nide* or *prit*) appropriately classified these children in 98 percent of instances. Gleitman and Rozin (1977) and Rozin and Gleitman (1977) taught failing readers and inner-city children with poor prognosis for reading acquisition a logography (morpheme script), a syllabic script, and an alphabetic script. This high risk population differed most from successful readers in acquiring the analytic alphabetic script, less for the syllabic script, and hardly at all for the morpheme script. The essential difficulty for poor readers seems to be in accessing their phonological machinery. They have the requisite phonological machinery in their heads, as Eimas showed: their problem is how to get to it.

The same point can be made about the historical sequence that led to the invention of alphabets, apparently. For the quite conscious business of designing a writing system, it seems easiest to think about language in terms of whole words. Thus morpheme scripts are invented earliest in historical time, and are separately reinvented in many cultures; syllabic scripts are invented much later and more rarely; while the alphabet was invented but once, and as late as the first millenium B.C. (Gelb, 1952).

Summarizing, tasks that involve conscious recognition and manipulation of linguistic units and classification of these are difficult for adults, and performance is variable. Skills in giving judgments also appear late in development. On the contrary, the perception and production of speech is carried out below the level of consciousness. In tasks that require perception, but not awareness, learning is fairly uniform over the human population and to some extent is already given at birth. Finally, where judgments are required, humans find it easiest to access high-level (or fully processed) linguistic representations, and much more difficult to process lower level (syntactic and phonological) representations.

*PARAPHRASES*

The findings just presented fit naturally with many demonstrations that people perceive the paraphrastic relations among certain sentences. More precisely, they perceive phrases and their paraphrases as equivalent, and hence do not distinguish them. For example, Bransford and Franks (1971)

presented a series of sentences to adult subjects. Later, they gave the
subjects partly different sentences, but which coded the same semantic notions
as the first set. The subjects could not recall which of the second series
they had actually heard on the earlier presentation. Thus while the gist
had been stored, the particular syntactic form had been lost to memory.
Sachs (1967) and Fillenbaum (1966) have reported similar effects. When time
passes, linguistic stimuli are unavailable for verbatim report but the semantic facts remain. Possibly, then, the relatively early or "raw" stages of
linguistic processing, during which phonological and syntactic facts are
perceived (and phrases are thus different from their paraphrases) decay
fairly rapidly; perhaps this is why they are relatively inaccessible to
reflection.

But while it is obvious from these demonstrations that paraphrastic
relations among sentences are perceived (or differences between paraphrases
erased from memory), it is not altogether easy to elicit judgments about
paraphrases in laboratory demonstrations.

Gleitman and Gleitman (1970) and Geer, Gleitman, and Gleitman (1972)
studied the abilities of adults to produce and recognize paraphrases of novel
nominal sequences (compound nouns). The stimuli in the experiments generally
consisted of sequences of three simple words. Two of the words were fixed
nouns such as *bird* and *house*. The third word was another noun (such as *foot*)
or a verb or adjective (such as *kill* or *black*); of course some of these words
had alternate categorial status (e.g., *kill* can be used nominally). The
three-word sequences were taped and presented orally, with either of two
stress patterns (132 or 213 stress) that are common for compound nouns. As
the words were combined in various orders, this procedure yielded some simple
nominal phrases such as *black bird-house* and *black-bird house*, but it also
yielded some sequences which are harder to interpret, such as *bird house-
black* and *bird-black house*. After suitable instructions, subjects were asked
to produce or recognize phrasal paraphrases of these sequences. That is,
the subject is being asked to realize that a *black-bird house* is a *house for
black-birds* or a *house where birds who are black live*. The task is, we
believe, a relatively natural and transparent one with which to inquire
whether people can think about the relatedness among sentences. Everybody
has been asked, from time to time, to say something "in his own words." Thus
it seems an easy matter, when asked for another phrase meaning the same as
*black bird-house*, to respond "That's a bird-house painted black." Surely

this is easy enough to do in the context of real conversation. Upon seeing a black bird-house, even for the first time, presumably an adult can say "Look: there's a black bird-house." No normal English-speaking adult would, we presume, say instead "Look! there's a bird-house black!"

Yet we found massive differences between two educational groups (clerical workers and Ph.D. candidates) in the ability to perform a variety of tasks related to paraphrasing compounds of this kind. On many occasions, the clerical workers *would* maintain that *bird-house black* was another way of saying *black bird-house,* contrary to what we believe their speech performance would be like. In fact, on a variety of paraphrasing tasks (even with simpler two-word compounds) there was no overlap at all in performance scores for members of the clerical and Ph.D. candidate groups. Figure 1 shows these population effects for a forced-choice task which required subjects to identify the correct paraphrases, from two choices, for three-word compound nouns.

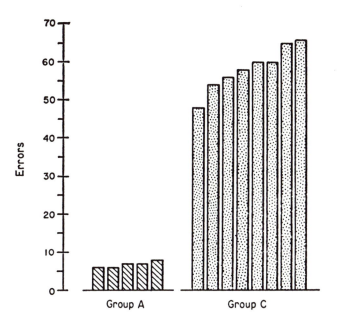

Figure 1. Forced-choice performance on a paraphrasing task by highly educated (A-group) and less educated (C-group) subjects. There were 144 stimuli and the less educated group choose wrongly in almost half. There was no overlap in the distribution of error scores by group (from Gleitman & Gleitman, 1970).

A closer look at the findings reveals that the group differences were larger or smaller depending on the particular oddity in the stimulus phrase: the group differences were largely attributable to syntactic, not semantic, problems posed by the novel compound nouns. Thus the two groups paraphrased more or less equivalently such semantic oddities a *house foot-bird* ("a bird with large feet who lives in houses," or "a live-in livery-bird"). On the contrary, only the most educated group handled perceived syntactic oddity by changing the categorial assignments of words (e.g., *bird house-black* was paraphrased by an educated subject as "a blackener of houses who is a bird" and *eat house-bird* as "a house-bird who is very eat."). The response style of the clerical group was quite different. These subjects approached syntactic oddities by ignoring, rather than manipulating, their syntactic properties. *Bird house-black* was typically paraphrased by this group as "a black bird who lives in the house"; *eat house-bird* was paraphrased as "everybody is eating up their pet birds." In short, when taxed, the average group focused on meaning and plausibility, while the highly educated group focused on the syntax even when meaningfulness was thereby obscured (as it surely was in the response "a house-bird who is very eat").

Notice that the syntactic oddities in these materials posed greater problems than the semantic oddities for both groups; but also that the between group differences were much greater for the syntactic oddities than for the semantic oddities. Manipulation and puzzle solving with low-level syntactic features seem to be attributes of linguistically talented people. The difference is apparent even in so far as one can show that the syntactic structures in question are handled adequately, in the context of normal speech and comprehension, by both populations.

Consider as an example Table 1 which lists the free paraphrases of both educational groups for the item *house-bird glass*. We can assume that every speaker of English, approximately, knows how to use *glass* both adjectivally (*a glass house*) and nominally (*a piece of glass; a glass to drink from*). Yet the less-educated subjects often interpreted *house-bird glass* as *glass house-bird*, a *house-bird made of glass*, or even as *glass bird-house*. Why not *glass used to make a house-bird* or *the glass used by the house-bird*, solutions which simultaneously resolve the semantic and syntactic properties of the stimulus item? (Notice particularly that there are no clear differences of semantic oddity for the two response types: anyone who can conceive of and believe in a glass house-bird ought to be able to conceive of and

Table 1

## PARAPHRASING HOUSE-BIRD GLASS

Responses of two populations to the task of paraphrasing the orally presented novel compound, *house-bird glass*. Hyphenation in the cited form represents the internal subcompound (i.e., the stress on this whole compound noun is 132). An asterisk marks the responses that fail to take into account the fact that the last word in such compounds is the head noun and thus must be the first (leftmost) noun in a paraphrase (a relative clause or preposi- tional phrase) that mirrors its syntactic and semantic properties. Thus the head noun of the compound is *glass*. For the internal sub-compound (*house- bird*), the same principle should apply: the rightmost noun (*bird*) in the com- pound is its head, and thus should appear in the leftmost position of a re- lative clause or prepositional phrase paraphrase of it. One Ph.D. candidate, but six clerical workers, err in applying this principle consistently, for this example. Similar performance disparities were observed for 144 similar stimuli, as well as in simplified (two-word) versions of them, and under a variety of task conditions (from Gleitman & Gleitman, 1970).

Responses of seven Ph.D. candidates:

1. glass for making house-birds
2. a very small drinking cup used by a canary
3. glass for house-birds
4. glass for house-birds
5. *a way of describing thickness of glass--glass as thick as (or in the shape of) glass of a bird-house
6. glass that protects house-birds
7. the glass that is produced by birds around the house

Responses of seven clerical workers:

1. *a glass house-bird
2. *house-bird that's in a glass
3. a drinking glass or a cup made out of glass of a bird in a house
4. *a bird that is made of glass
5. *a special glass to use in a bird's house
6. *a house-bird made from glass
7. *a glass house-bird

believe in the glass which is used to manufacture such house-birds.) But even in a forced-choice situation, when both options were displayed, the clerical group still preferred the inversion. The structure of these findings suggests that only the most-educated group will consider least-common categorial assignments for the component words of the stimuli (i.e., *glass* as noun rather than adjective) in this situation.

In short, the clearest difference between these populations is in *focusing* on the syntactic issues, accessing and manipulating language knowledge in a non-communicative setting. Clearly this does not imply that adults are all equal in their ability to analyze complicated meanings in everyday life. But across a range broad enough to be of considerable psychological interest, all normal individuals can realize consciously that some expressions within their semantic compass (however limited this may be) are "meaningless" or "odd in meaning." Everyone realizes that there is something peculiar about the sentence *George frightened the color green* and can "fix it up" via some semantic change. But not everyone can focus on a syntactic anomaly and perform an appropriate syntactic manipulation to repair it, even if they are in productive control of the construction during ongoing conversational exchange. In this sense, "meaning" can be brought to conscious attention more readily than can syntactic form.

One final property of subjects' functioning in the paraphrase task is of special relevance to the study of ambiguity detection we will report in the third section. Disruption of morphological boundaries in the stimulus phrases had a massive effect on subjects' performance, regardless of group membership. Many of the stimuli involved the sequence of words *bird, house* and many others involved *house, bird*. But apparently the compound *bird-house* (a house in which birds live) is represented holistically, at least for certain purposes, by English speakers, as a single unit with morphological integrity; i.e., a bird-house is just a cage. On the contrary, the compound *house-bird* (a pet bird that lives in the house), while equally plausible (by analogy, e.g., to *house-cat*) is uniformly interpreted as morphologically complex--having two pieces, being a compound, not a single word.

We draw this distinction on the following evidence: if the stress pattern in the stimulus three-word compound required that the subsequence *bird, house* be taken as a linked sub-unit, subjects' performance was especially good. Thus *bird-house black* was easier to paraphrase than *house-bird black*. (Whatever differences of sheer plausibility may contribute to differences in response adequacy here, these wash out over the mass of stimuli.) But if the stress pattern of the compound was such as to violate the unity of *bird-house* or *house-bird*, then the former became harder than the latter, i.e., *bird house-black* was harder to paraphrase correctly than *house bird-black*. This interaction of familiarity (prior lexical knowledge of *bird-house* but not *house-bird*, despite semantic plausibility of both) with "unity" (preservation or violation of the integrity of the pair, by shifts in stress pattern) is shown in Figure 2.

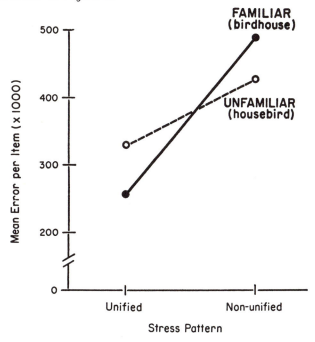

Figure 2. The effect of morpheme familiarity under conditions of unified and non-unified stress on error scores of adults paraphrasing compound nouns (from Gleitman & Gleitman, 1970).

Reasonably enough, the "errors" that were made were responses that ignored the stress pattern of the stimulus, with the effect of honoring the morphological integrity of *bird-house*. Thus the subjects were inclined to paraphrase *bird house-black* as "Paint used on bird-houses," but less inclined to paraphrase *house bird-black* as "Paint used on house-birds." This was so despite the fact that the taped stimuli had been pretested for detectability of the stress patterns. (Note that the highly-educated group correctly paraphrased even those stimuli showing unit disruption of this sort most of the time. The effect of the "unity" dimension was in terms of the same relative increase of error, regardless of group membership; but still there was a huge difference, by group, in the absolute frequency of error for these stimuli.)

Summarizing, subjects in this task, regardless of group membership, were affected massively in their response characteristics by (1) the syntactic properties of the stimulus phrase and (2) the lexical or morphological boundary properties of the phrases. Hence, overall, these experiments displayed the enormous effects of syntactic and morphological patterning in human language performance. But moreover, the subject populations differed radically in their response styles. While the most-educated group was biased to attend to surface syntactic properties of the stimulus when taxed, the least-educated group was biased to attend to the plausible semantic interpretations of the separate words in the stimulus, under the same circumstances. We conclude that the ability to manipulate superficial levels of language structure in non-communicative settings is a property of linguistically able (or experienced) individuals.

*AMBIGUITY*

One of the mysteries of language perception is the relative inobtrusiveness of the massive ambiguity in the sound stream. Hardly a sentence of any complexity fails to yield at least some stretched ambiguity of sound, syntax, or meaning, and yet we are usually unaware of this. In fact, it is rare that we notice anything but the meaning the speaker had in mind to express. MacKay (1966) asked listeners to complete sentence fragments that were or were not ambiguous. While alternate potential interpretations had clear effects on processing, in terms of lengthening subjects' response times, the subjects almost never reported noticing any ambiguities. This was so despite the fact that MacKay chose stimuli for which the two interpretations were roughly equipotent. Curiously, these subjects did stutter or giggle very often when the stimuli were ambiguous (a metalinguistic itch, perhaps) but yet were unaware of ambiguities as they performed the task.

Of course, as is the case for paraphrase and classification, adults can provide conscious judgments of ambiguity, at least if the stimulus materials are not too difficult. MacKay and Bever (1967) were successful in getting adult subjects to detect and report on lexical, surface, and deep structure ambiguities in sentences; however, their experiment was not designed to tap the limits of judgmental skills, but rather to measure certain processing capacities in their subjects. To determine whether subjects' reflective access to language structure for ambiguous sentences is in line with these skills in the context of classification and paraphrase tasks, it would be necessary to make the task harder, either in terms of the linguistic materials, or by choosing subjects who are less able in ways relevant to the task. One obvious first step is to examine children's reactions to linguistic ambiguity, since we already have evidence that the young child is weak in judgmental performance.

There are a number of relevant studies in the literature. For example, Shultz and Pilon (1973) asked children to report on phonological, lexical, surface, and deep structure ambiguities, using paraphrase (explication of both readings) as the measure. They report that phonological and lexical ambiguities were detected and adequately paraphrased by younger children (in the range from 6 to 9 years) than were syntactic ambiguities. Only 50 percent of the syntactic ambiguities were adequately described by their 15 year-old subjects.

In part, these findings are in accord with what we ought to predict: lexical ambiguities, which turn on transparent aspects of the semantics of whole-word units, are more accessible to reflection than rapidly decaying syntactic representations of sentences. But for one thing, the delay in developmental onset is much greater than we might anticipate. After all, even the five-year old subjects of Gleitman, Gleitman and Shipley (1972) gave some judgments and paraphrases in quite difficult syntactic circumstances. We reserve this issue for somewhat later discussion. But it also comes as a surprise to us, given our position and the findings in the reading literature, that phonological ambiguities were among the easiest for Shultz and Pilon's subjects to detect and explicate.

However, the results of this investigation are not altogether easy to interpret, owing to the classification of the stimulus types. Primarily, the problem is that these authors call the ambiguity of *He often goes to the bank* (river or financial institution) "lexical," but they call the ambiguity of *He saw three pairs (pears)* "phonological." Apparently they had in mind the spelling distinction, or possibly a postulated level of underlying phonological structure (or historical relationship among roots) on which such spelling distinctions might be explicable. But surely this distinction is unrealistic for first-graders (i.e., for speaker-listeners not in control of the Latinate vocabulary, and hence whose phonological representations are not so complexly related to phonetics; for discussion, see Chomsky, 1970; Moskowitz, 1973; Gleitman & Rozin, 1977). Moreover, the phonological ambiguities classed together in this study include some that involve actual sound differences (*line* vs. *lion*) and some that do not (*patients* vs. *patience*). And they include some that involve restructuring of morpheme boundaries (e.g., *eight tea cups* vs. *eighty cups*) and some that do not (e.g., *paw* vs. *pa*). We know, partly from studies cited earlier, that such factors significantly affect subjects' response styles. Therefore the phonology vs. lexicon findings by Shultz and Pilon require some further experimental review.

There is another study of the emergence of judgments of ambiguity that bears centrally on issues in metalinguistic capacity. Kessel (1970) asked children to interpret orally presented ambiguous sentences. His subjects ranged in age from 6 to 13 years. Though his older subjects did better than Shultz and Pilon's, they were still failing to see through to certain ambiguities 10 to 15 percent of the time, even at 12 and 13 years of age. (Of course, there is every reason to suppose that some ambiguities, even with type equated, are harder than others and that some subjects are more facile than others; hence the rough equivalence in findings of the two studies is all we can expect.) Furthermore, again as we would predict from the discussion that has preceded, subjects' success in the task was partly a function of linguistic level of the ambiguity in the stimulus sentence. While 6 and 7 year olds comprehended both meanings of most lexical ambiguities (e.g., the distinction between *pipe* and *pipe*), underlying ambiguities (e.g., *She hit the man with the glasses*) came under control in the period from 8 to 9 years (i.e., error rate showed the sharpest drop in this period), and surface structure ambiguities (e.g., *He told her baby stories*) came under control in the period from 9 to 12 years. That is, the more superficial levels of language representation were the more difficult to reflect and report on. (Note that Kessel provided two oral presentations in case the alternate interpretations were associated with different intonation contours.)

Perhaps most interestingly of all, Kessel reports a qualitative change in subjects' response style at about 12 years:

> "within those children detecting the ambiguities, a marked difference
> exists between the fifth graders and all [younger] children in what
> might be termed explicit awareness of the ambiguities. Thus for the

surface structure ambiguities, the fifth graders spontaneously made comments such as "You can put it differently ..." (Kessel, p. 45).

Finally, it is worth noting that while Kessel found growth in metalinguistic skill to be a function of age, there was much individual variation among his subjects: some 7 year olds could do what other 10 year olds could not. Similar findings of individual difference in explicating ambiguity have been reported elsewhere (Fowles & Glanz, 1977; Brodzinsky, 1977). This outcome accords with the supposition (Gleitman & Gleitman, 1970) that there are considerable individual differences in metalinguistic capacity.

Given the finding of late appearance of skills with ambiguity, there is some impetus to seek better inquiry procedures: perhaps there are conditions under which children might display further skill with this kind of task. After all, the classification task used by Gleitman, Gleitman, and Shipley (1972) was highly motivating: the subjects laughed uproariously on hearing each anomalous sentence (e.g., *Golf played my sister*) and eagerly "explained" them all.

Many investigators have hit upon the verbal joke as a relatively natural and familiar medium, one which might maximally reveal whatever competence children have in the domain of ambiguity recognition (Brodzinsky, 1977; Fowles & Glanz, 1977; Shultz & Horibe, 1974). As we noted in introductory comments, American children seem to be well versed in the structural formats for riddles and jokes, and apparently the task is highly motivating. Moreover, the joke form is disingenuously designed to handle the problem of interpretive bias. Let us digress here to review this issue, and ask how it is manipulated in a verbal joke.

For certain of their stimuli, MacKay and Bever (1967) found that almost all their subjects would report a particular reading first. Presumably, given the wording of the stimulus sentence and plausibilities in the world, listeners are sensitized toward one interpretation over another potential one--the second is accessed only by brute force, if the task demands it. Then if one potential interpretation was reported first, say, by 18 out of 20 of their subjects, MacKay and Bever called the item "biased." But if the order of report over subjects was about 50-50, the item was termed "unbiased."

Presumably, we understand each other as well as we do, despite the potential structural ambiguity in discourse, because we exploit such biases, given the topics of conversation and likely states of affairs in the world.

But then one might wonder, symmetrically, how people "get" ambiguity jokes
as easily as they do. At least part of the answer seems to be that contextual
cues are used in jokes, first to bias the listener powerfully in one direction
in the "set-up," and then to bias him in the other direction in the "punch-
line." Hence the listener has been forced to veer from one interpretation
to the other, in succession (for some discussion of this isse, see Fowles
& Glanz, 1977). For shrouded reasons, the interpretative switch seems to
arouse listeners; it seems to be a component of humor.

Let us consider an example. In the set-up, a genie appears from a lamp,
prepared to do service:

Genie: I am your slave and will grant your every wish.
Man  : Well, while I'm thinking of a really important wish, *make me a
milkshake*.

Clearly, the listener (unless he is a genie) is biased toward benefactive
interpretation: *Make a milkshake for me*. But then the punch line:

Genie: Poof! You're a milkshake!

requires re-viewing the original sentence as *Make me into a milkshake,* or
*Make a milkshake out of me*, in which case the genie is malefactive indeed.
The bias shift is a component of the humor here (though, no doubt, so is
the "funny visual picture," the consternation of the master confounded and
the slave triumphant, the category error, i.e., while man can conceive being
transformed to frog or even giraffe, it is strange to mutate to a milkshake).

Evidently, bias shift is a property of certain verbal jokes, which makes
their ambiguity more perspicuous to the listener. Accordingly, many investigators have presented ambiguities to young children in the context of jokes
and riddles.[1] And indeed subjects give evidence of competence at earlier
ages in this setting. Fowles and Glanz (1977) found appreciation of syntactic humor in some second and third graders, and Brodzinsky (1977), who
studied nine-year olds, similarly found that many of them could unravel
riddles that turned on syntactic ambiguities.

In summary, the structure of development for judging ambiguity seems
similar to that for classifying and judging paraphrase. There is relatively
late appearance of the skill, significant individual variation, and sensitivity in response style to the language level that is being tapped. Especially
delayed onset for judgments of ambiguity, compared to classification or

paraphrase, seems to be the result of interpretive biases for particular sentences and situations. This bias seems to be at least partly overcome in the verbal joke format, which thus seems an ideal context in which to look at the competence of young children in making judgments of ambiguity. We report now on one study of the issues here.

AMBIGUITY JUDGMENTS

Reported here is a study of children's ability to explain jokes that turn on various kinds of language ambiguity.

*STIMULI*

Jokes were culled from children's joke books and magazines. They were pretested for vocabulary difficulty, using as subjects a heterogenous group of first graders. Thirty jokes whose vocabulary was understood by these six-year olds were selected for use in the experiment. They were recorded by two young actors (a male and a female); one actor spoke the set-up of the joke, and the other delivered the punch-line. The jokes were delivered with intonation pattern as neutral as possible between the two interpretations. This was fairly easy to do for ambiguities not involving phonological distortion, but hard to do in the phonological instances (i.e., nobody knows how to say something between *b* and *p* so as neutrally to represent either *barking* or *parking*). Hence, for phonological instances, the punch-line was delivered against the bias of the set-up; that is, if the set-up prepared the listener to hear *parking* the punch-line was delivered as *barking*. The jokes were classified according to ambiguity type, and presented in a randomized order. This order was counterbalanced by simply reversing the joke order (half the subjects heard one order, the other half the other). A constraint on the randomization procedure was that instances of a single joke type not be exhausted in a very few presentations. The classification was as follows (see Table 2 for two instances of each type):

A) *Phonological:* An ambiguity that results when two similar phonetic sequences (which differ only in a single phonological segment) identify two separate words, which have different meanings, e.g., *cracker/quacker*. Six jokes turning on this problem were used.

B) *Lexical:* An ambiguity that results when a single phonological sequence identifies two separate words, which have different meanings, e.g., *bark/bark*. Seven jokes turning on this problem were used.

C) *Surface structure:* An ambiguity that results when a single sequence of words can be bracketed in two different ways, identifying different sentential meanings, e.g., *(man) (eating fish) / (man eating) (fish)*. Four jokes turning on this problem were used.

D) *Underlying structure:* An ambiguity that results when a single sequence of words has two transformational sources, or two case labelings, identifying different sentential meanings, e.g., *make me a milkshake* as *make a milkshake for me / out of me*. Four jokes turning on this problem were used.

E) *Morpheme boundary:* An ambiguity that results when a polysyllable can be interpreted as a single morpheme or as a sequence of morphemes, e.g., *engineers / engine ears*. Five jokes turning on this problem were used.

F) *Morpheme boundary with phonological distortion:* An ambiguity that results from the interaction of a phonological problem (as in type A) acting together with a morpheme boundary problem (as in type E) e.g., *let's hope / let's soap*. Four jokes turning on this complex problem were used.

The use of differing numbers of jokes of each type was an unprincipled consequence of the difficulty of finding good instances of each, that met the criteria of simple wording, etc.

*SUBJECTS*

The subjects were 48 children from a predominantly white upper middle-class school in the suburbs of Philadelphia. Four boys and four girls were selected from each of grades 1-6; thus there were 24 girls and 24 boys in the sample, and 8 children in each of six grades. Two of the boys and two of the girls from each grade were chosen on the basis of their "very good" reading ability, by estimate of the reading specialist in the school, and the other two pairs were chosen on the basis of their "very poor" reading ability by the same estimate. Thus there were 24 good readers and 24 bad readers in the sample. But note that "good" or "bad" reader in this middle-class population is not, realistically, good and bad by American norms. The readers called "good" at this school are way above national norms, and probably the bad readers are close to the national average. Thus only rarely (we guess) were any of these subjects at the lower end of any sensible scale of verbal talent or achievement.

Each child met with the experimenter individually, in an unoccupied classroom. The instructions were read and the first joke played from the recording. The subject rated the joke as "very funny," "a little funny," or "not funny at all," and then tried to explain it. Then the next joke was played from the recording, etc. The sessions were taped on a separate recorder. If the subject wanted the joke re-presented, this was allowed.

## INSTRUCTIONS

"I hear that you like jokes. Is that true?" (if "yes"): "Good." (if "no"): "Well I have some good ones. Would you like to hear them?" (if "no," the subject was rejected, and another one found). "I have some jokes on this tape recorder. Some people have told me that some of them are really good and that some of them are really bad. I cannot tell which are the good ones and which are the bad ones. I need your help. Would you listen to the joke and tell me if you think it is very funny, a little funny, or not funny at all?...." (at this point the subject is given an example joke; in fact, probably the worst joke of all time--the one about the chicken crossing the road; almost all subjects responded "not funny" without hesitation). "OK, now how would you explain this joke to someone who doesn't get it?" (after the subject responds, one more example is given; this time, the joke is not as bad). "OK? Do you understand how we do it? Remember, this is not a test, I really need your help. Are you ready...?"

## SCORING

The child's explanations of the jokes were scored on a five-point scale:
1 - Got the joke immediately and explained both sides of the ambiguity.
2 - Got the joke, but only after encouragement to try harder.
3 - The same, but only after yet more encouragement to try harder still.
4 - The same, but only after a push of the sort: "He said 'so and so,' right? Are there two meanings for 'so and so'?"
5 - Did not get the joke, but knew the meaning of all the words in the joke, and both meanings for the ambiguous words (as determined by further inquiry).

Table 2

SAMPLE JOKES, CLASSIFIED IN TERMS OF THE SOURCE OF AMBIGUITY

| | |
|---|---|
| A: Phonological | 1. If you put three ducks in a box what do you have? A box of quackers. |
| | 2. Bob coughed until his face turned blue. Was he choking? No, he was serious. |
| B: Lexical | 1. How can hunters in the woods best find their lost dogs? By putting their ears to a tree and listening to the bark. |
| | 2. How do we know there was fruit on Noah's ark? Because the animals came in pairs. |
| C: Surface Structure | 1. How would you run over a dinosaur? I'd start at his tail, run up his back, then over his neck and I'd jump off. |
| | 2. Where would you go to see a man-eating fish? A seafood restaurant. |
| D: Deep Structure | 1. We're going to have my grandmother for Thanksgiving. You are? Well, we're going to have a turkey. |
| | 2. Will you join me in a bowl of soup? Do you think there's room for both of us? |
| E: Morpheme Boundary -No Phonological Distortion | 1. Why can one never starve in the desert? Because of the sand which is there. |
| | 2. How do trains hear? Through their engine ears. |
| F: Morpheme Boundary with Distortion | 1. Do you think that if I wash, my face will be clean? Let's soap for the best. |
| | 2. Did you read in the newspaper about the man who ate six dozen pancakes at one sitting? No--how waffle. |

If the subject did not give evidence that he knew the meanings of all the words in the joke, that item was removed from analysis for that subject. Instances of this sort were limited to a small percentage of jokes for the first-grade poor readers. Given the purposes of the experiment (to determine whether, under optimal circumstances, these subjects can talk about two interpretations of a linguistic event) only the distinction between the score 5 and all other scores (1 through 4) was considered relevant, and this is the distinction on which all analyses were based.[2] The subject's "funniness rating" of the joke, and also the experimenter's estimate of the subject's "mirth response" were also noted for each joke. These measures were taken because they have some plausible connection to the general issues of humor appreciation as it is related to ambiguity detection. However, findings here are of no centrality for the judgmental questions we have been discussing, to our knowledge (see Hirsh-Pasek, forthcoming, for discussion of the findings in terms of children's humor) and so are not reported here.[3]

## RESULTS

All analyses were based on the subject's percent errors of explication: here, the percentage of the subject's responses that were failures to explain the ambiguity in the joke even when prompted (type 5 responses). These error scores were analyzed in terms of the subjects' sex, grade (= age) and reader group, as well as the stimulus variables of presentation order and, most important, ambiguity type. There was no effect of presentation order. On the other hand, females outperformed males at each grade level, though this result is of no particular interest in the present context. Presentation order and sex were therefore ignored for purposes of further analysis.

## The Effect of Age and Reading Ability

Figure 3 shows the distribution of error scores for each of the six joke types, plotted against grade for both good and poor readers. Not surprisingly, performance is better for older than for younger children and for superior readers over less adequate readers. These effects are highly reliable. A two-way analysis of variance yielded a significant main effect of grade ($F = 6.36$, $df = 5$ and $36$, $p < .01$) and of reading ability ($F = 28.87$, $df = 1$ and $36$, $p < .001$) and no interaction ($F < 1$, $df = 5$ and $36$, $p < .05$).

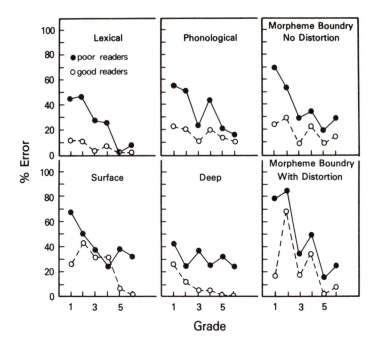

<u>Figure 3.</u>  Mean percent errors in reporting ambiguities by good and bad readers, by grade.

*The Analysis of Ambiguity Effects*

One of our main questions in this inquiry concerned the relative difficulty of the different joke types.  Our first task was to find analyses that could tease apart some of the factors that contributed to their difficulty.

Two separate analyses were performed toward this end.  The first compared jokes that turn on more superficial syntactic representations (Type C) with those whose point hinges on semantically more relevant representations (Type D).  This was done by means of a three-way analysis of variance, with grade (1-6), reader group (poor vs. superior) and syntactic type (Type C vs. Type D; see Table 2) as the three factors.  All calculations were based on the mean percent error for each subject and each joke type.

A second analysis considered the effect of disrupting morpheme boundaries; that is, the "unity effect."  For this purpose, we compared jokes which preserved morpheme boundaries on both interpretations (that is, lexical and phonological ambiguities; Types B and A) to jokes which changed these boundaries across interpretations (that is, morpheme boundary ambiguities which did or did not involve phonological distortions; Types F and E). To undertake this comparison, a four-way analysis of variance was performed. The four factors were grade, reading ability, presence or absence of phonological distortion, and presence or absence of morpheme boundary ambiguity. The comparison between the morpheme-boundary and the phonological distortion effects is schematized in the two-by-two table below:

|  | - Unity disruption | + Unity disruption |
|---|---|---|
| - phonological distortion | Type B: lexical ambiguities | Type E: morpheme boundary ambiguities without phonological distortion |
| + phonological distortion | Type A: phonological distortion ambiguities | Type F: morpheme ambiguities with phonological distortion |

*Ambiguity Effects: General Findings*

Both analyses of variance showed main effects of grade (F = 2.44, df = 5 and 36, p < .10; F = 7.95, df = 5 and 36, p < .01) and reading ability (F = 14.00, df = 1 and 36, p < .001; F = 28.81, df = 1 and 36, p < .001). There were no interactions with any of the ambiguity types. As a result, it seems reasonable to consider performance collapsed across grades and reading groups. The mean percent error obtained in each of the six joke types is shown in Table 3.

As the table shows, some kinds of jokes are evidently easier to explain than others. Easiest of all are ambiguities that turn on transparent semantic properties of the stimulus sentences: ambiguities of lexical interpretation, and underlying structure or case-labelling ambiguities. Harder are ambiguities that hinge on superficial representations of the sentence: phonological ambiguities and surface structure ambiguities. Special problems are posed by disruptions of unit boundaries, especially if these are compounded by phonological distortions.

Table 3 - Experiment on ambiguity judgments.  On the left is the type of
ambiguity played on in jokes, on the right the overall percent errors made.

| | | |
|---|---|---|
| Type A: | Phonological (*joking* vs. *choking*) | 27% |
| Type B: | Lexical (*bark* vs. *bark*) | 16% |
| Type C: | Surface structure (*man-eating fish* vs. *man eating fish*) | 33% |
| Type D: | Underlying structure (*grandmother for Thanksgiving: to eat* or *to eat with*) | 21% |
| Type E: | Morpheme boundary (*engineer* vs. *engine ear*) | 30% |
| Type F: | Morpheme boundary with phonological distortion (*let's soap* vs. *let's hope*) | 39% |

*Ambiguity Effects: Syntax*

One of our analyses compares the contributions of underlying and superficial syntactic ambiguities to task difficulty.  We have argued that ambiguities that depend on surface representations (Type C) should pose greater
difficulty for judgment than ambiguities that depend on underlying representations (Type D); and Table 3 shows that the surface structure jokes were
harder to get than the underlying structure jokes.  The difference of difficulty in the expected direction proved to be highly reliable (F = 11.56,
df = 1 and 108, p < .001).  As already noted, similar findings have been
reported by Kessel (1970) and others.


*Ambiguity Effects: The Role of Phonological Distortion*

We have previously argued that studies such as that of Shultz and Pilon
(1973) confounded two possible determinants of performance in the explication
of ambiguity: phonological distortion and morpheme-boundary effects.  As Table
3 shows, both of these variables seem to matter.  When the joke hinges on a
phonological distortion, it is harder to get.  The same holds for cases in
which the ambiguity restructures morpheme boundaries.

Let us first look at the effect of phonological distortion.  Our second
analysis of variance (grade x reading ability x presence or absence of phonological distortion x morpheme boundary effects) showed a highly reliable
effect of phonological distortion.  The F-ratio was 11.04 with df of 1 and
108 and p < .001.

It is reasonable to ask why we obtained a considerably greater effect
of phonological distortions than such investigators as Shultz and Pilon,
whose subjects seemed to find phonological ambiguities among the easiest of
all to report on.  On the face of it, their outcomes seem to violate our ex-

pectations. Surely phonological representations are "raw," or relatively "meaningless" or relatively "early in processing" and so (on our story) should be hard to contemplate. We discussed earlier that the findings for phonological ambiguity are often contaminated by problems of stimulus classification, e.g., Shultz and Pilon called the contrast *sale / sail* a phonological ambiguity. But even more important is the issue of stimulus presentation for phonological, compared to other ambiguities.

As already noted, it is not possible to give a phonetically neutral rendition of most phonological ambiguities. If a joke turns on the relation of *joking* to *choking*, one must say "ch" or "j" in the punch-line, for there is no articulation in between that we can make. On the contrary, if a joke hinges on the contrast of ((*man eating*) (*fish*)) / ((*man*) (*eating fish*)), it is relatively easy to provide a neutral rendition. For *bank / bank* or *grandmother for Thanksgiving*, no alternative intonation patternings are even possible. Which rendition should be offered in the punch-line for the phonological ambiguity?

The fact is, if the wrongly biased version (the same one as in the set-up context) is provided, approximately nobody gets the jokes (e.g., *Why are ducks like cookies? Because they are animal crackers!*) Therefore, the appropriately biased version is typically provided in a punch-line (*Because they are animal quackers!*). That is, an unheralded phonological ambiguity is hard indeed. We conclude that phonological distortions in jokes are rendered artificially easy because they are given away by special phonological information in the punch-line; if the bias is appreciated, even isolated sentence presentations of ambiguity will tend to emphasize the least-biased phonological form. Since we made such feeble effort as we could to neutralize the give-away in our punch-lines (by special instructions to our actors) we reinflated the difficulty of phonological distortion. But we could not go all the way (provide wrongly biased punch-lines) if subjects were ever to respond correctly. Hence, as Table 3 shows, the phonological ambiguities are not hardest of all in our findings, though probably they "really" are hardest.

*Ambiguity Effects: The Role of Unity*

We previously noted that speakers in paraphrasing tasks had much more difficulty if they were forced to consider alternate morpheme boundary analyses of sentences. *Bird-house black* is relatively easy to paraphrase; *bird*

*house-black* is murderously difficult. The trouble is not simply the issue
of conceptualizing what kind of house-black goes naturally with birds. An
additional problem is posed by the fact that *bird-house* forms a lexical unit
which must be forced apart if the stress pattern requires a *house-black*
linkage. The same problem is posed by the necessity to view *engineers* as
*engine ears*. This resegmentation is no mean feat for youngsters, even with
the semantic hints provided by bias shift in the punch-line. Flexibility
of analysis at this level is as difficult with ambiguities as it is with
paraphrase tasks. This difficulty is shown by the fact that Type E and F
jokes together led to a greater error percentage than did jokes of Type A
and B together (34.5% vs. 21.5%). This difference was highly significant,
as shown by the analysis of variance ($F = 18.77$, $df = 1$ and $36$, $p < .001$).
There was no interaction between this factor and the effect of phonological
distortion nor with grade or reading ability. (Similarly, in the paraphrasing
task of Gleitman & Gleitman, unity disruptions increased error rate to the
same extent for all populations.)

*Verbal Talent and the Ability to Explicate Jokes*

We have commented on the highly significant effect of reading ability
on error scores (see Figure 3). The size and stability of this effect is of
some interest. As we remarked earlier, the skills of speaking and understanding do not seem to vary very dramatically across ranges of intelligence
(though, to be sure, there is a paucity of hard information on this topic,
aside from the special findings for retarded individuals). Yet there is a
large and consistent difference in performance on tasks we take to be metalinguistic, as a function of such talent differences. We earlier documented
the same population differences in paraphrasing tasks with adults. We seem
to see them also in terms of the level of adequacy of subjects in the two
reader groups with ambiguity problems, at each age level. Possibly also of
interest is the fact that the effects of verbal talent are large and clear
on both formal analyses of these data, while the effects are smaller and
less stable in terms of subject age. Such a finding of differential potency
of talent and age variables in predicting metalinguistic performance would
fit in with the suggestions by Kessel (1970) and Fowles and Glanz (1977)
that there are age-independent significant effects of verbal talent in reflective capacities with ambiguity.

In all candor, there is a further effect we expected to document, but did not. Specifically, we expected that the performance adequacy of good readers and bad readers would be quite similar, with age equated, for ambiguities that turned on semantic properties of the stimuli, but that greater performance disparities would be found for ambiguities that turned on superficial representations. Such outcomes would bolster the interpretations that come to mind from our work on paraphrase and reading: that differences in syntactic, not semantic, dimensions account for the bulk of differences between more and less able judgment-givers. These effects ought to appear, in the present experiment, as interactions between ambiguity type and reader group. No such interactions were found. There are, of course, many ways to explain away such noneffects. Most important, as described above, probably few of our subjects were actually "below average" in verbal talent or achievement. Thus the sample is highly attenuated for this variable, which we know is an important and relevant one for metalanguage performance. But still we are at least sobered by this null outcome.

*SUMMARY OF THE RESULTS*

Along with other investigators, we have found that the ability to explain ambiguity emerges much later in developmental time than the ability to detect potential alternative interpretations of a single speech signal. The six- and seven-year old subjects rarely provided adequate responses in terms of a dimension related to ambiguity, while the ten- and eleven-year old subjects performed adequately. Response adequacy was clearly related to verbal talent, measured grossly in terms of reading skill. The poor readers performed consistently at a level about a year behind the good readers.

More interestingly, adequacy with the task was a function of certain linguistic variables that were manipulated in the stimulus materials. First, superficial syntactic and phonological ambiguities were harder to explain than ambiguities closer to the level of meaning. Moreover, just as in paraphrasing tasks described by Gleitman and Gleitman (1970), subjects had special difficulty when the task required breaking up what is evidently an indivisible language unit, the polysyllable morpheme.

SUMMARY AND CONCLUSIONS

For tasks such as classifying sentences, producing paraphrases, and explaining ambiguities, conscious reflection on language "as an object"

evidently is required. On the contrary, speaking and understanding require
no such conscious reflection. The reflective skills appear later in life
and show more individual variation than do skills related to language perception and production. At one end of the spectrum of conscious capacity
with language stand those young children who can utter and detect phonological
identities, e.g., *pet* either as an independent morpheme or as a syllable embedded in *carpet*: and yet give no evidence of *knowing that they know* about
the phonological identity. Here is an example from a six-year old good
reader:

Joke : "Did you ever stand on a pet?"

    "Stand on a pet? I should say not!"

    "I have: on a carpet!"

Subject: Well, he said have you ever stamped on a pet--I never would.

Exp. : You wouldn't? Well, what kind of pet did he stand on?

Subject: He said I haven't stepped on a rug.

Exp. : Did he say *rug?*

Subject: Uh-huh.

Exp. : I don't think I heard that.

Subject: I think I did.

Sometime later in development, subjects seem to hear the meaning in the stimulus both ways (though we cannot be entirely sure, for any one case) and yet
the judgment escapes them; here is an example from a seven-year old good
reader:

Joke : "What did one hat say to the other hat?"

    "You stay here; I'll go on ahead"

Subject: Not funny, cause hat can't move except when they are on your
    head.

We suspect the child perceived the sense go-on-ahead (why else did he say
"move?"). But also he seems to have caught go-on-a-head (why else did he
say "they are on your head?"). But even if so, surely explicit awareness
of the ambiguity is faint; this is the same beguiling inadequacy, Kessel
(1970) reported in some of his subjects. The tantalizing partial insight,
the agonizing miss, is even closer in the following sequence from a nine-
year old good reader. Possibly this time the child really has understood
the ambiguity. But either it flits in and out of his head so he cannot
manipulate it; or else, perhaps, he has the idea that he must make the ambiguity fit the imaginary situation in both its interpretations at once.

In that case, this subject is displaying awareness of ambiguity, but failure to see how such ambiguities are to be exploited in verbal jokes:

Joke    : "My father makes faces all day."
          "Why does he do that?"
          "He works in a clock factory."

Subject: I can see. You can make clocks. You can make a face of a clock but you can't make the face of a clock on a face.

Exp.   : What did he mean by 'making faces'?

Subject: He makes faces all day.

Exp.   : Show me what you mean.

Subject: Making faces like this (Child makes a funny face)

Exp.   : Oh

Subject: ... on the clock

Exp.   : Oh, making faces like that on the clock

Subject: He was making faces like that on the clock.

In sum, what must logically be there in the brain for the purposes of perceiving language is not necessarily always or wholly there in the mind for the purpose of reflecting on and commenting about language. The subject does not know *about* everything he knows *how to do* with language. If this is so, we might begin to speculate concerning the performance disparities in linguistic and metalinguistic tasks. Possibly, the brain makes available to consciousness only some of the computations relevant to language perception. Our findings would suggest that those structural representations that must be computed early in sentence processing decay rapidly and are not transferred to consciousness. On the contrary, the representations of language that are to be stored and manipulated in the service of meaning-representation decay less rapidly, and leave traces in consciousness. Thus they are available for reflection.

On this suggestion, it might be that linguists asking speaker-listeners for judgments on classification, ambiguity, and paraphrase, would biasedly receive information about the more meaningful levels of representation. But the psycholinguist who measures language responses in the first milliseconds (and who notoriously mistrusts outcomes whose response latencies are long and variable) would receive his most reliable data from earlier stages of processing. Discrepant results from these two domains of inquiry might be partly understandable on this difference in the reliability and validity of data at some particular level of analysis. To the extent that this might be true, it may not be valuable to ask who has the corner on psychological reality.

However, there are many other possible interpretations of the obvious discord between language judgments and language use. Gleitman, Gleitman, and Shipley (1972) speculated that the metacognitive function may be a single example of a more general "metacognitive" organization in humans. That is, a variety of cognitive processes seem themselves to be the objects of higher-order cognitive processes in the same domain. Examples of metacognition in memory would be recollection (when we know that we remember) and intentional learning (when we know we must store the material for later retrieval). On this view, there need be no formal resemblance between metacognition and the cognitive processes it sometimes guides and organizes. Rather, one might expect to find resemblances among the higher-order processes themselves. On this view, judgments (and therefore grammars) have little direct relevance to speech and comprehension, but rather to reasoning. Whatever resemblance exists between language processing strategies and grammars may derive from the fact that the human builds his grammar out of his observation of regularities in his speech and comprehension. Whatever differences exist between these organizations may derive from the fact that the reflective capacities have properties of their own, which enter into the form of the grammars they construct.

As of now, there is little concrete evidence to support either of these speculations against the other, or against a variety of alternatives. At best they are metaphors that give a framework for holding together our studies of language judgment. Returning to the level on which we have at least preliminary data: from the study of ambiguity and the context into which it seems to fit, we cannot answer exactly what the brain says to the mind. Surely it does not tell everything it knows, but quite conceivably remarks to the mind, "Oh, what a wit you are!"

*ACKNOWLEDGEMENTS*

We wish to thank the staff and students of the Bryn Mawr Elementary School and the University City New School for their participation in this study. Claire Gleitman and Nicki Silvers are thanked for their talented performance as joke tellers. Douglas Lax and Linda Garvin are thanked for transcribing and analyzing some of the interview sessions. Marion Gilbert is thanked especially for administrative support and astonishing mental fortitude in preparing this manuscript. We gratefully acknowledge the support of the William T. Carter Foundation, which allowed us to do this work; and the Max-Planck-Gesellschaft, Projektgruppe für Psycholinguistik, whose conference on metalinguistics and language learning, under the direction of Professor Pim Levelt, allowed us to share ideas with many interesting investigators in this field. Finally, we thank Ann Premack for stylistic suggestions which we think lightened the presentation of this material.

*FOOTNOTES*

1. Most of the investigators we have been citing were primarily interested in the far more subtle and general facts about humor appreciation, and its growth in the young child; the problem of ambiguity detection was thus usually a mere vehicle in these studies, though that does not diminish their relevance to the metalinguistic issues we are discussing here.

2. However, the stratified scoring may have some interest of its own, and results described in this form are available from the authors on request.

3. Again, the outcomes on these other scorings are available from the authors on request.

*REFERENCES*

Blank, M. Cognitive processes in auditory discrimination in normal and retarded readers. *Child Development*, 1968, 39, 1091-1101.

Bransford, J.D., & Franks, J.J. The abstraction of linguistic ideas. *Cognitive Psychology*, 1971, 2, 331-350.

Bresnan, J. A realistic transformational grammar. In: M. Halle, J. Bresnan, & G. Miller (eds.), *Linguistic theory and psychological reality*. M.I.T. Press, in press.

Brodzinsky, D.M. Children's comprehension and appreciation of verbal jokes in relation to conceptual tempo. *Child Development*, 1977, 48, 960-967.

Brown, R. *A first language: The early stages*. Cambridge, Mass.: Harvard University Press, 1973.

Chomsky, N. *Aspects of the theory of syntax*. Cambridge, Mass.: The M.I.T. Press, 1965.

Chomsky, N. Phonology and reading. In: H. Levin & J.P. Williams (eds.), *Basic studies on reading*. New York: Basic Books, 1970.

Cutting, J.E., & Rosner, B.S. Categories and boundaries in speech and music. *Perception and Psychophysics*, 1974, 16, 564-570.

Downing, J. (ed.), *Comparative reading: Cross-national studies of behavior and processes in reading and writing*. New York: Macmillan, 1973.

Downing, J., & Oliver, P. The child's conception of "a word." *Reading Research Quarterly*, 1973-1974, 9, 568-582.

Eimas, P.D., Siqueland, E.R., Jusczyk, P., & Vigorito, J. Speech perception in infants. *Science*, 1971, 171, 303-306.

El'konin, D.B. (Translated from the Russian by R. Raeder & J. Downing). In: J. Downing (ed.), *Comparative reading: Cross-national studies of behavior and processes in reading and writing*. New York: Macmillan, 1973.

Feldman, H., Goldin-Meadow, S., & Gleitman, L. Beyond Herodotus: The language-like communication of linguistically deprived deaf children. To appear in: A. Lock (ed.), *Action, gesture and symbol*. New York: Academic Press.

Fillenbaum, S. Memory for gist: Some relevant variables. *Language and Speech*, 1966, 9, 217-227.

Firth, U. *Components of reading disability*. Doctoral dissertation, University of New South Wales, 1972.

Fodor, J.A., Bever, T.G., & Garrett, M.F. *The psychology of language: An introduction to psycholinguistics and generative grammar*. New York: McGraw Hill, 1974.

Fowles, B., & Glanz, E. Competence and talent in verbal riddle comprehension. *Journal of Child Language*, 1977, 4, 433-452.

Geer, S.E., Gleitman, H., & Gleitman, L. Paraphrasing and remembering compound words. *Journal of Verbal Learning and Verbal Behavior*, 1972, 11, 348-355.

Gelb, I.J. *A study of writing: The foundation of grammatology*. Chicago: University of Chicago Press, 1952.

Gleitman, H., & Gleitman, L.R. Language use and language judgment. In: C. Fillmore & W. Wang (eds.), *Individual differences in language ability and language behavior*. New York: Academic Press, in press.

Gleitman, L.R., & Gleitman, H. *Phrase and paraphrase*. New York: W.W. Norton & Co., 1970.

Gleitman, L.R., Gleitman, H., & Shipley, E. The emergence of the child as grammarian. *Cognition*, 1972, 1, 137-164.

Gleitman, L.R., & Rozin, P. The structure and acquisition of reading. I. Relations between orthographies and the structure of language. In: A.S. Reber & D. Scarborough (eds.), *Toward a psychology of reading*. Hillsdale, N.J.: Erlbaum, 1977.

Holden, M.H., & MacGinitie, W.H. Children's conceptions of word boundaries in speech and print. *Journal of Educational Psychology*, 1972, 63, 551-557.

Kessel, F.S. The role of syntax in children's comprehension from ages six to twelve. *Monographs of the Society for Research in Child Development*, 1970, 35 (6).

Kuhl, P.K., & Miller, J.D. Speech perception by the chinchilla: Voiced-voiceless distinction in alveolar plosive consonants. *Science*, 1975,

190, 69-72.

Lackner, J.R. A developmental study of language behavior in retarded children. In: D.M. Morehead & A.E. Morehead (eds.), *Normal and deficient child language*. Baltimore: University Park Press, 1976.

Lenneberg, E.H. *Biological foundations of language*. New York: Wiley, 1967.

Liberman, I.Y., Shankweiler, D., Fischer, F.W., & Carter, B. Explicit syllable and phoneme segmentation in the young child. *Journal of Experimental Child Psychology*, 1974, 18, 201-212.

Maclay, H., & Sleater, M. Responses to language: judgments of grammaticalness. *International Journal of American Linguistics*, 1960, 26, 275-282.

MacKay, D.G. To end ambiguous sentences. *Perception and Psychophysics*, 1966, 1, 426-436.

MacKay, D.G., & Bever, T. In search of ambiguity. *Perception and Psychophysics*, 1967, 2, 193-198.

Marks, L., & Miller, G.A. The role of semantic and syntactic constraints in the memorization of English sentences. *Journal of Verbal Learning and Verbal Behavior*, 1964, 3, 1-5.

Miller, G.A. Some psychological rules of grammar. *American Psychologist*, 1962, 17, 748-762.

Miller, G.A., & Isard, S. Some perceptual consequences of linguistic rules. *Journal of Verbal Learning and Verbal Behavior*, 1963, 2, 217-228.

Morehead, D.M., & Ingram, D. The development of base syntax in normal and linguistically deviant children. In: D.M. Morehead & A.E. Morehead (eds.), *Normal and deficient child language*. Baltimore: University Park Press, 1976.

Moskowitz, B.A. On the status of vowel shift in English. In: T.E. Moore (ed.), *Cognitive development and the acquisition of language*. New York: Academic Press, 1973.

Newport, E., Gleitman, H., & Gleitman, L.R. Mother, I'd rather do it myself: Some effects and non-effects of maternal speech style. In: C.E. Snow & C.A. Ferguson (eds.), *Talking to children: Language input and acquisition*. Cambridge: Cambridge University Press, 1977.

Pylyshyn, Z.W. The role of competence theories in cognitive psychology. *Journal of Psycholinguistic Research*, 1973, 2, 21-50.

Rosner, J. Auditory analysis training with prereaders. *The Reading Teacher*, 1974, 27, 379-384.

Rosner, J., & Simon, D.P. *The auditory analysis test: An initial report*. University of Pittsburgh: Learning Research and Development Center, Publication 1971/3, 1971.

Rozin, P., & Gleitman, L.R. The structure and acquisition of reading. II. The reading process and the acquisition of the alphabetic principle. In: A.S. Reber & D. Scarborough (eds.), *Toward a psychology of reading*. Hillsdale, N.J.: Erlbaum, 1977.

Rozin, P., Poritsky, S., & Sotsky, R. American children with reading problems can easily learn to read English represented by Chinese characters. *Science*, 1971, 171, 1264-1267.

Sachs, J.S.  Recognition memory for syntactic and semantic aspects of connected discourse.  *Perception and Psychophysics*, 1967, 2, 437-442.

Shatz, M.  Semantic and syntactic factors in children's judgment of sentences.  Unpublished manuscript, University of Pennsylvania, 1972.

Shultz, T., & Horibe, F.  Development of the appreciation of verbal jokes.
*Developmental Psychology*, 1974, 10, 13-20.

Shultz, T., & Pilon, R.  Development of the ability to detect linguistic
ambiguity.  *Child Development*, 1973, 44, 728-733.

Singer, H.  IQ is and is not related to reading.  In: S. Wanat (ed.), *Intelligence and reading*.  Newark, Del.: International Reading Association, 1974.

Slobin, D.  Cognitive prerequisites for the development of grammar.  In:
C.A. Ferguson & D.I. Slobin (eds.), *Studies of child language development*.  New York: Holt, Rinehart & Winston, 1973.

Slobin, D.  The more its changes--on understanding language by watching
it move through time.  In: *Papers and Reports on Child Language Development*, No. 10.  Stanford University, Department of Linguistics, September,
1975.

Smith, D.M.  Creolization and language ontogeny: A preliminary paradigm for
comparing language socialization and language acculturation.  In: C.
J.W. Bailey & R.W. Shuy (eds.), *New ways of analyzing variation in
English*.  Washington: Georgetown University Press, 1973.

Wanner, E.  Review of J.A. Fodor, T.G. Bever, & M.F. Garrett, *The psychology
of language*.  *Journal of Psycholinguistic Research*, 1977, 6, 261-269.

Wepman, J.M.  *Wepman auditory discrimination test*.  Chicago: Language Research
Associates, 1958.

# Levels of Semantic Knowledge in Children's Use of Connectives

G.B. Flores d'Arcais

Psychology Laboratory, University of Leiden, Leiden, The Netherlands
and
Max-Planck-Gesellschaft, Projektgruppe für Psycholinguistik
Berg en Dalseweg 79, Nijmegen, The Netherlands

This paper deals with children's knowledge of some simple relational terms in Dutch and Italian. The words in question are mainly the common connectives used to introduce clauses of reason or cause, of purpose, of result, and of time. Acquisition of the meaning of these words was studied here in several different experimental contexts. On the basis of the results, it will be argued that different strata of semantic competence exist, and that access to some levels of this knowledge emerges earlier than to others. However, we will not attempt to examine or explain the ontogeny of the child's notions of the conceptual relations which are expressed through these words. This problem goes far beyond the limits of the present paper. Although semantic knowledge for the connectives studied and awareness of the usual relations implied by these words may well be interdependent, our primary interest here was children's increasing access to the words' meaning. As Vygotsky (1962, p. 46) observed, "the child may operate with subordinate clauses, with words like *because, if, when,* and *but,* long before he really grasps causal, conditional or temporal relations themselves."

A child's (or adult's) grasp of the meaning of some term $X$ can be defined in various ways: one may define it as understanding $X$ as part of a sentence one comprehends; as recognizing that in that context $X$ can or cannot be replaced by another word without the meaning of the expression being changed; as realizing that $X$ is more (or less) appropriate than other words in a given context; as grouping *X* with words similar in meaning and not with others; and so on.  In fact, the kind of tacit knowledge needed to perform different tasks such as comprehending in context, judging meaning equivalence, choosing appropriate descriptions, and classifying according to meaning may be quite different.  The latter tasks, though perhaps more abstract and metalinguistic in character, seem to demand greater awareness of words' meaning proper, apart from the speech contexts in which they are used.  We will see that children of a certain age do quite well on some of these meaning-relevant tasks and quite poorly on others.  The data collected suggest progressive awareness of the stability of meaning across contexts, and finally awareness of meaning out of context.  In line with this, it will be argued that the acquisition of meaning is a continuous development, and not an all-or-none process.

The paper is organized as follows.  First, the results of some experiments on children's comprehension of two-clause sentences joined by the connectives are summarized.  Five other studies are then presented. These latter experiments dealt with judgments of meaning equivalence between such sentences, preferences for them as descriptions of simple picture sequences and narrative stories, and judgments of meaning similarity among connectives presented in isolation.

## COMPREHENSION OF COMPLEX SENTENCES BY CHILDREN

The following results emerged from a series of experiments on children's recall and comprehension of sentences having largely to do with cause and effect, and composed of a main and subordinate clause joined by a conjunction. The experiments, done in Dutch and Italian, are reported in detail elsewhere (Flores d'Arcais, 1978a, 1978b; Flores d'Arcais, Joustra, & Joustra-De Boer, 1976).

1.  For sentence recall (subjects were Italian children aged six to nine), children showed a clear tendency to reproduce subordinate clauses formed with *perché* ('because') and *cosicché* ('so that') as coordinate: the cause tended to be recalled in first position, the effect in second position. When reproductions retained subordinate/main structure, subordinate clauses of cause (formed with *perché* 'because') tended to be recalled in first position, and subordinate clauses of purpose (formed with *affinché* 'in order to') in second position. That is, subjects showed a tendency to organize the two events according to cause-effect order.

2.  In a comprehension study (subjects were Italian children aged 3;2

to 8;6), younger children tended to choose according to an "order-of-mention strategy": in cause- and purpose-naming sentences what is mentioned first tends to be carried out first, what is mentioned second to be executed second. Later the "order-of-mention strategy" was still used, but the older children also seemed to show some understanding of the subordinator. By the age of seven to eight most children appear capable of understanding the sentences correctly without much difficulty.

3. In a comprehension study of causal sentences (subjects were Dutch children aged three to seven), we found that sentences with the order main/ subordinate (effect/cause) were significantly more difficult than those with the order subordinate/main, when order of mention is the same as cause-effect order; and, secondly, when children only carry out one action, at all ages they more often carry out the action expressed in the main clause (the effect).

One can seriously question, however, whether correctly *acting out* such sentences with toys (the classical technique used here) really implies full understanding of them, including understanding of all the "important" lexical items present. There are several possible answers to this question with respect to the relational words at issue. The first possibility is that the child's comprehension in the experiments above does indicate full understanding of the connectives. The semantic knowledge of the child would then correspond approximately to that of the adult. In that case, we would expect the child to show some understanding of the connectives in other contexts and in isolation as well.

Another possibility is that correctly performing the experimental task does not *per se* imply that the meaning of the different subordinating conjunctions is fully understood. In most causal sentences that a child is exposed to (and in most of the sentences used in the experiments) the causal relation can probably be inferred without specifically processing the connective. If we say, for example, "The glass broke because it fell off the table," it is perfectly possible to understand the sentence even if one does not understand the connective *because*. The information present in the two clauses is such that the relation of physical causality that is expressed is the most likely one in real life. Possibly, the child understands such complex sentences correctly, but his comprehension is largely based on the information given in the two clauses. In this case, we would not expect the child to understand the connective *because* when a finer or more complex distinction is called for.

A third possibility compatible with the results referred to above is that, as well as understanding the sentences correctly, the child has some notion of the meaning of the connectives themselves. When the connectives are presented in isolation, however, his level of semantic knowledge is still insufficient for an adequate performance. That is, the child's knowledge of the meaning of the connectives would be partial; he might process them adequately in certain contexts, but his knowledge would be very limited and different from an adult's knowledge.

In the experiments reported below we have tried to find out how children of different ages perform in different tasks involving connectives that are similar or identical to those used in the experiments described above. The general purpose of the experiments was simply to find out when children are able to correctly perform the more demanding tasks, for example, to make judgments of equivalence of meaning for two sentences that are identical except for their connective. Such tasks we believe require a more abstract knowledge and use of the meaning of the critical words.

## JUDGMENTS OF CORRESPONDENCE IN MEANING

The first and second experiments reported below deal with children's judgments of meaning correspondence between connectives when these are presented in complex sentences. Given pairs of sentences that were identical except for the connective, subjects were asked to judge whether the sentences meant the same thing. Thus, children were in effect asked if they were aware of differences in meaning between the connectives, and of their possible "interchangeability" in a given context. When two sentences were judged as non-equivalent, the child was credited with making a semantic distinction between the two connectives involved.

### Experiment 1

### Judgments of Equivalence of Meaning by Italian Children

The experimental material consisted of eight pairs of Italian sentences, identical except for the connectives. Two pairs of connectives were used: *perché* ('because') vs. *affinché* ('in order to'); and *prima che* ('before') vs. *cosicché* ('so that'). As additional material, two-picture sets representing simple events were used to create some context for the sentences. For example, the first picture of one set showed a vase falling off a table, the second a

vase lying broken on the ground. The "appropriate" sentence describing the
two pictures was the first one of the following pair:

    (a) Il vaso si è rotto perchè è caduto per terra ('The vase broke *because*
it fell on the ground');

    (b) Il vaso si è rotto affinchè cadesse per terra ('The vase broke *in
order* (for it) *to* fall on the ground').

The pictures and sentences were embedded in other material of a similar type
to avoid "set" effects. In 12 of 16 such filler items, both sentences were
appropriate for describing the pictures shown. The sentences were written
on cards, one sentence per card.

    Sixteen Italian children each from grades two and four (mean age 7;8
and 9;7 respectively) took part in the experiment. Each child was tested
individually. For each test item, the child was given the set of two pictures
and presented with the card with the "appropriate" sentence in it. The child
was then shown the second sentence (the "incorrect" one) and asked to decide
whether he would be willing to use it to tell a friend about the event expressed in the first sentence and represented in the pictures. The two sentences shown to the child always contained one of the pairs of connectives
or the other.

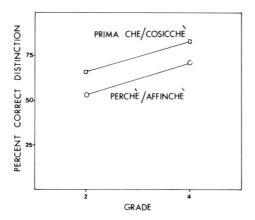

Figure 1. Experiment 1. Percent inappropriate sentences rejected as picture
descriptions.

*Results*. The percentage of refusals for the two pairs of connectives is shown in Figure 1. There is the expected increase with age in appropriate distinctions. The distinction between cause-stating *perché* and purpose-stating *affinché* appears to be less accessible or less salient at both age levels than the distinction between neutral but consequent-stating *prima che* and result-stating *cosicché*.

### Experiment 2
### Judgments of Equivalence of Meaning by Dutch Children

The experimental materials were two-clause Dutch sentences with the structure main clause-connective-subordinate clause. The following connectives were used: *omdat* ('because'); *zodat* ('so that, such that'); *doordat* ('since, because of the fact that') and *voordat* ('before'). Three sentences were constructed with each of the four connectives. Then, from the three original sentences with each connective, three further sentences were constructed. The "new" sentences were identical to the original ones except for the connective, which was replaced by the ones remaining. This gave a total of 24 sentences in 12 pairs, each pair consisting of an original sentence and a sentence identical to it except for the change in connective. For example for sentence (a) formed with *omdat*, a second sentence (b) was constructed with *zodat*:

(a) De hond blaft, omdat de poes eraan komt ('The dog barks, *because* the cat is approaching');

(b) De hond blaft, zodat de poes eraan komt ('The dog barks, *so that* the cat approaches').

Each sentence was typed on a separate card. The subjects in the experiment were 20 Dutch children from the following grades: grade 1 (approximate age 7 years); grade 2 (8 years); grade 4 (10 years) and grade 6 (12 years). The children were tested individually. For each trial, they were given a card with a sentence on it (always one of the 12 original ones), and were required to read it aloud. Immediately after this they were given the corresponding sentence with the connective replaced, and were asked to decide whether they could use this sentence to tell another child about the event described in the first sentence. Unlike Experiment 1, no pictures were used. Before the experiment, four pairs of synonymous sentences (e.g., "The mother tells a story" vs. "The mother tells a tale") were presented to familiarize the child with the task of judging equivalence of meaning between sentences.

The 12 pairs were then presented in random order with four "dummy" pairs inserted in the experimental material to avoid "set" effects.

*Results.* Responses were the judgments for each pair of sentences. Frequencies and percentages of these judgments were computed for each age group and for each pair of sentences. Here and in the other experiments, the differences found were tested with the simple parametric test for significant differences in proportions, both between groups of Ss and within groups of Ss. The results are shown in Figure 2(a)-(d). The connective in the original sentences is shown at the top. Each subfigure reports the percentage of judgments of equivalence given for the sentences with the three substituted connectives with respect to this. So, for example, Figure 2(a) gives the number of times (expressed in percentages) sentences containing the connectives *doordat, voordat* and *zodat* were judged as equivalent to the sentences first shown to the children with *omdat*. Age groups are shown at the bottom of each figure. Obviously, the lower the percentage, the better the distinction for the pair involved is established. In Figure 2(a) the curve for *zodat*, for example, indicates that grade 6 children rarely accepted a sentence with *zodat* as equivalent to the "same" sentence with *omdat*.

In the first and second grades, the percentage of judgments of equivalence is rather high for most of the pairs of connectives. With the exception of *omdat* ('because') and *doordat* ('since, because of the fact that') which are largely synonymous in this case and serve as a kind of control, there is a continuing decrease in judgments of equivalence after the first grade, with a clear "drop" between the second and fourth grade. A clear differentiation begins to emerge only in the fourth grade. For almost all the pairs of connectives (the exceptions are *voordat* 'before' equivalent to *zodat* 'so that,' and *doordat* to *voordat*) the differences shown in percentages of judgments between grade two and grade four are significant. This decrease may be taken as indicative of increased sensitivity to the meaning of the individual connectives.

Notice that one should expect a kind of symmetry among the results. If a sentence with connective X is judged as non-equivalent to the same sentence with connective Y, then at the same age level the converse should also hold, namely a sentence with Y should be judged non-equivalent to the comparable sentence with X. In general, this is the case. However, the proportions of judgments sometimes show rather striking differences. At grade four, 76%

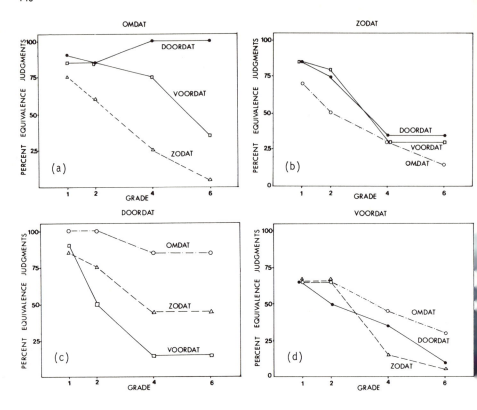

Figure 2. Experiment 2. Percent sentences judged equivalent when the connective at top is substituted for by the three other shown.

of Ss judged sentences with *voordat* as equivalent to sentences with *omdat*, but only 48% judged sentences with *omdat* as equivalent to sentences with *voordat*. This difference might be due simply to "sampling error" in the selection of sentences. Or, as an alternative explanation, the organization of semantic knowledge concerning these terms may not be bi-directional. Rather, it may be analogous to the case of associations and retrieval of synonyms which are not always purely symetrical: if B is the primary associate of A, A is not always the first associate of B.

Finally, a few remarks can be made about the differentiation of particular pairs. The connective *voordat* seems to be most clearly distinguished from all others even at the lower age levels (compare Figure 2(d) with the others). *Zodat* begins to be clearly distinguished from the other three connectives at the fourth grade level. *Doordat* is differentiated more easily and sooner from *voordat* than from *zodat*, while related *omdat* is differentiated more easily from *zodat* than from *voordat*. If we look at the overall results, we can suggest the following progressive differentiation in the knowledge of the meaning of the connectives. The pairs which are differentiated most easily and earliest are *doordat/voordat* and *omdat/zodat*. At the fourth grade, the distinction *voordat/zodat* is also very well established. The differentiation of *omdat/voordat* comes more slowly.

*PREFERENCE FOR A SENTENCE AS A DESCRIPTION OR A PARAPHRASE*

The next two experiments deal with children's choice of one out of a small set of sentences, as a description of a relation presented through pictures, or as a paraphrase of a brief story. The sentences in each set were identical except for the connectives. Only one of the sentences was an appropriate paraphrase of the story. Choice of sentences with inappropriate connectives was taken as an indication of "confusion" among the meaning of the words, while reduction in the number of incorrect sentence choices for a particular connective was taken as an indication of its progressive differentiation from the others.

*Experiment 3*
*Choice of Sentences Explaining Picture Stories*

Sets of two pictures were used to represent a relation between two simple events. For each set of pictures, three Italian sentences identical in structure except for the connective were constructed. The following Italian connectives were used: *perché* ('because') *affinché* ('in order to') and *cosicché* ('so that'). Only one sentence was an appropriate description of each set of pictures, while the other two were entirely inappropriate.

The relations chosen were: two events, where the first picture depicts an event which is *explained* by the event in the second picture (Type 1); two events where the first event is the *cause*, and the second the *physical consequence* or *result* (Type 2). For example, the following two pictures were used for a Type 1 relation. First picture: *A man runs to the station*. Second picture: *The train is already leaving; the man looks exhausted and disappointed*. The sentences used with these pictures were:

(a) L'uomo corre alla stazione perché il treno sta partendo ('The man runs to the station *because* the train is leaving') (The appropriate description).

(b) L'uomo corre alla stazione affinché il treno stia partendo ('The man runs to the station *for* the train *to* leave').

(c) L'uomo corre alla stazione cosicché il treno stia partendo ('The man runs to the station *so that* the train leaves').

Notice that in Italian the second clause always has the same structure-- unlike the English translation given here--except that (a) is indicative and (b) and (c) are subjunctive.

For the Type 2 relation, pictures of the following scenes are an example. First picture: *A boy eats a big ice-cream.* Second picture: *The boy has a stomach ache.* The sentences in this case were:

(a) Il bambino mangia troppo gelato, cosicché si ammala ('The boy eats too much ice-cream, *so that* he gets sick') (The appropriate description).

(b) Il bambino mangia troppo gelato, affinché si ammali ('The boy eats too much ice-cream, *in order to* get sick').

(c) Il bambino mangia troppo gelato, perché si ammala ('The boy eats too much ice-cream, *because* he gets sick').

The experimental material consisted of three sets of Type 1 relations and of three sets of Type 2 relations and their corresponding sentences. The materials were presented along with six other picture and sentence sets corresponding to other types of relations. Each sentence was written on a card.

Eighteen Italian children each from grades two and four (mean age 7;7 and 9;10 respectively) took part in the experiment. The children were tested individually. A child was shown one set of pictures and was asked to explain "what was going on" in the pictures. He was then presented with the three cards with the sentences on them in random order. The experimenter read the sentences aloud and asked the subject to choose the one which was the most appropriate for the events shown in the pictures. In each case, only one sentence was appropriate, namely the sentence with *perché* for a Type 1 set and the sentence with *cosicché* for a Type 2 set.

*Results.* The proportion of choices of the different connectives are represented in Figure 3(a) and (b). The figures report both the appropriate choices (sentences with *perché* for the first type of material and with *cosicché* for the second type), and the inappropriate choices. At both age levels *perché* ('because') is chosen correctly slightly more frequently than *cosicché* ('so that') (the difference is significant, $p < .05$). *Affinché* ('in order to') is in both cases confounded more often with both *perché* and *cosicché* than they are with each other.

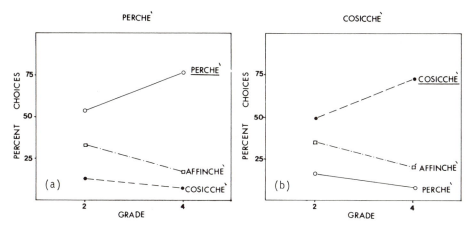

<u>Figure 3.</u> Experiment 3. Percent choices of Italian sentences describing picture sets, with the appropriate sentence's connective shown at the top.

*Experiment 4*

*Choices of Sentences Explaining Written Stories*

Nine "basic" sentences, all with main clause-subordinate clause structure, were used in this experiment. For each sentence three versions were formed, one with *omdat* ('because'), one with *voordat* ('before'), and one with *zodat* ('so that, such that'). Each sentence was typed on a card. For each triplet of sentences a short story was prepared, in which a clear causal, temporal or consecutive relation between the events was expressed.
For example:

"The dog is sitting in the garden. At a certain moment, a cat arrives. The dog sees the cat. He then begins to bark."

The three sentences used with this story were:

(a) De hond blaft, omdat de kat eraan komt ('The dog barks, *because* the cat arrives').

(b) De hond blaft, voordat de kat eraan komt ('The dog barks, *before* the cat arrives').

(c) De hond blaft, zodat de kat eraan komt ('The dog barks, *so that* the cat arrives').

As is apparent, only one of the three sentences, namely (a), is an appropriate

"explanation" of the last part of the "story." For all the "stories" and
the sets of corresponding sentences presented, only one of the sentences was
an appropriate description.

The subjects were 20 Dutch children in each of the following age groups:
grade two (mean age 8;2); grade four (mean age 10;2); and grade six (mean
age 12;1). The children were tested individually. The nine stories, preceded
by two training items, were presented to each child in random order. After
each story, the three cards with the three related sentences were put in
random order on the table in front of the child. The experimenter read the
three sentences and while doing so pointed at the corresponding cards. The
subjects' task was to choose the sentence which expressed "the same meaning"
or "was an appropriate explanation" for the story they had just heard.

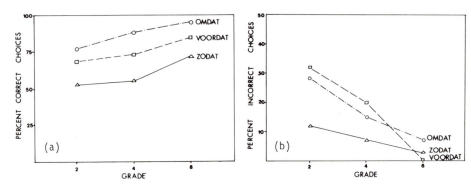

Figure 4. Experiment 4. Percent correct (a) and incorrect (b) choices of
sentences formed with three Dutch connectives to paraphrase short descriptions.

*Results.* Percentages of correct choices are given in Figure 4(a), and
incorrect choices in Figure 4(b). For sake of clarity, Figure 4(b) does not
include double choices: the percentages reported in the two figures therefore do not sum to unity. In this experiment, *omdat* ('because') seems the
most easily and earliest differentiated, while *zodat* ('so that') is the most
difficult for the child. At all three age levels, *omdat* was correctly chosen
the most frequently, and *zodat* the least frequently. The differences are in
each case significant between the .05 and .01 levels.

As for the incorrect choices, *omdat* and *voordat* ('before') were also
chosen significantly more frequently ($p < .05$) than *zodat* at grades two and
four; *omdat* was chosen more frequently than the other two connectives at the
grade six level. The pattern of incorrect choices thus seems to reflect the

pattern of correct ones: the order of preference is the same, namely *omdat*, *voordat*, and then *zodat*. A tentative explanation is that when the child does not know the meaning of the sentence, he chooses the one with the connective the meaning of which he knows best, or perhaps with which he is more familiar. (See, however, also Figure 2).

*JUDGMENTS OF SIMILARITY IN MEANING BY SORTING*

*Experiment 5*

Two experiments were actually carried out: a preliminary study with 20 third and sixth grade Dutch children and 15 connectives, and a more complete study with 20 subjects from each of the following age groups: grade two (mean age 8;1); grade four (mean age 10;1); grade six (mean age 11;11) and adults. In this second study the 20 most frequent subordinate conjunctions in Dutch (Uit den Boogaart, 1975) were used. As the results of the first experiment are rather similar to those of the second, only the results of the latter will be reported in detail. The Dutch connectives used in this study are shown in Table 1 with their closest English equivalents. The 20 connectives were typed on separate cards and subjects were instructed to sort the cards into as many piles as they wanted, putting together those words which they felt to be very similar in meaning.

Table 1 - Experiment 5. The 20 most frequent Dutch subordinate conjunctions used for the sorting task, together with their English equivalents. Though common in written text, a few of these words are only occasionally used in spoken discourse.

| Dutch | English | Dutch | English |
|-------|---------|-------|---------|
| *als* | 'as, if' | *ofschoon* | 'although' |
| *dan* | 'then' | *omdat* | 'because' |
| *daar* | 'as, because' | *tenzij* | 'unless' |
| *dat* | 'that' | *terwijl* | 'while, as' |
| *doordat* | 'since, because of the fact that' | *toen* | 'when, then' |
| *eer* | 'before' | *totdat* | 'until' |
| *hoewel* | 'although' | *voordat* | 'before' |
| *indien* | 'if, in case' | *wanneer* | 'when' |
| *mits* | 'provided that' | *zoals* | 'as, like' |
| *nadat* | 'after' | *zodat* | 'so that, such that' |

146

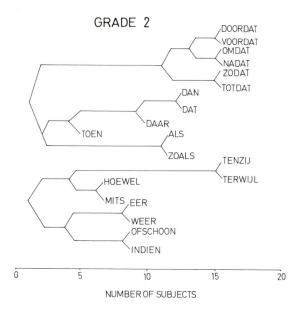

GRADE 2

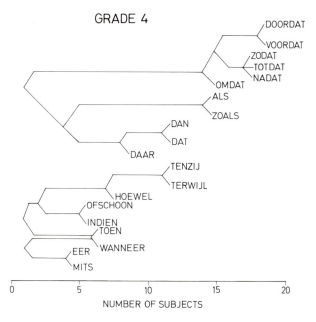

GRADE 4

Figure 5(a). Experiment 5. Hierarchial clustering structure for 20 Dutch connectives classified for meaning similarity by second and fourth grade children.

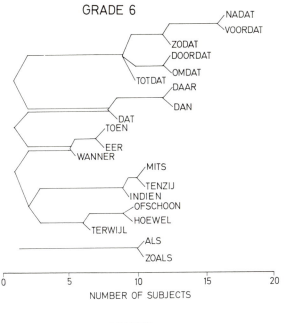

GRADE 6

NUMBER OF SUBJECTS

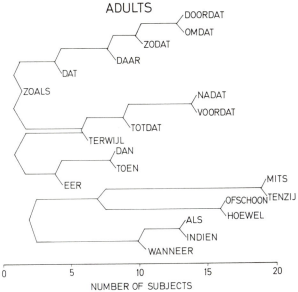

ADULTS

NUMBER OF SUBJECTS

Figure 5(b). Experiment 5. Hierarchial clustering structure for 20 Dutch connectives classified for meaning similarity by sixth grade children and adults.

*Results.* For each subject, each of the N (N-1)/2 pairs of connectives was given a score of 1 if the two had been put by the subject in the same pile and a score of 0 if the two were put in different piles. For each group of Ss, then, a "similarity" matrix was obtained, with entries which could vary from 0 to 20 for every pair of connectives--20 if all the Ss put two particular connectives in the same pile, 0 if none had. The matrices were treated using two procedures, Kruskal's multidimensional scaling method (1964) and Johnson's hierarchial clustering schema (1967). Only the results of the second analysis are reported here, and are shown by age group in Figure 5.

At the second grade level, the selection seems entirely based on phonetic and graphic similarity. All words ending in *–dat* are clustered together. The words *daar, dan,* and *dat,* of about the same length and beginning with the same letter, are clustered together. *Terwijl* and *tenzij* form another cluster, and *wanneer* and *eer* are put together, presumably because of the similarity of their last syllables.

In the fourth grade the structure obtained is still very similar to the one of the second graders. All connectives ending in *–dat* are again in the same cluster, and so are the short words beginning with *d*, the words with initial *t* and final vowel *ij*, and words ending in *als*. However, for certain pairs of words, the cluster structure also begins to reflect other classification criteria: for example, *wanneer* and *toen*, both temporal connectives, appear now in the same cluster.

At the sixth grade level, the structure is still much the same, but several other distinctions are made. For example, within the cluster of the words ending in *–dat*, the two temporal conjunctions *voordat* and *nadat* now appear in the same cluster, and so do the causal connectives *omdat* and *doordat* (which at the lower ages were more frequently paired with the physically similar *voordat*). The conditionals *mits* and *tenzij* now appear clustered together, and are correctly associated at a higher clustering level with *indien*.

Finally, the structure of the adult's responses reveals a clear distinction among the connectives, with hierarchies of clusters showing a fuller understanding of the differences in meaning and of the semantic categories involved. The three major clusters might be roughly labelled as *causal, temporal,* and *conditional*.

*Discussion.* The data obtained seem clear and straightforward. Two types of principles seem, in general, to underlie the classification. The first is

physical similarity of the stimulus words, including phonetic similarity
(initial syllable, final syllable, initial or final sound) and graphic similarity (spelling, word contour and word length). Knowledge of the meaning
of the connectives is the second principle underlying the classification.

The children at the lowest age level seem to take into account only
physical similarity. Developmentally, this principle gradually combines
with the principle of semantic similarity, but grade six children still
seem strongly influenced by the first principle. The first semantic distinction which seems to emerge, at the fourth grade level, concerns some temporal
conjunctions. By the sixth grade, there seem to be certain distinctions
within the class of conditionals as well. The general picture, however, is
of a mixed strategy. On the one hand, all the words ending in *-dat* are
clustered together. But within this cluster some semantic principles seem
to operate. Results from our earlier experiment suggest that this mixed
picture reflects the fact that some sixth grade children sorted the words
exclusively on the basis of physical similarity, others did so only on the
basis of semantic criteria, and a third group was guided by both principles
at the same time.

It is important to stress that the results cannot be explained as due to
failure in comprehending the instructions, or misapprehension of the task as a
request for classifications based on similarity of physical form. In the preliminary sorting experiment, the Ss were given different classes of words as
separate tasks (the stimili were 15 connectives in one case, 15 concrete nouns
in another and 15 adjectives in a third). The structure with the nouns was
very clear even with the youngest children: the choices were made on the basis
of semantic or pragmatic criteria, not on physical characteristics such as
words' length. At the age of eight or nine, then, children seem capable of
performing the sorting task appropriately with words from other domains.
This is consistent with the results of other investigations on the acquisition
of word meanings using this technique (e.g., Chapman, 1976). At the third
and fourth grade levels, Anglin (1970) found that children already operate
on the basis of clear semantic criteria--although their semantic representation may be rather different from the structure of the semantic domain in
adult subjects. As to the connectives, the knowledge needed to perform the
sorting task seems to develop much more slowly than, for example, with concrete nouns and adjectives. This development is evidently still in progress
at the sixth grade level.

*GENERAL DISCUSSION*

Let us recapitulate the main points made in this paper. In a previous series of experiments the developmental course of the comprehension of different types of complex sentences was examined (Flores d'Arcais, 1978a, 1978b; Flores d'Arcais, et al., 1976). One of the results of these experiments was that by the age of eight to nine years most children perform correctly when acting out complex sentences made up of two clauses joined by a connective. We credited the children with comprehension of the sentence structures involved.

The first two of the new experiments reported in this paper involved judgments of correspondence of meaning between complex sentences that are identical except for the connectives present. The third and fourth experiments involved the choice, among triplets of sentences identical except for the connectives, of one which expressed a certain meaning, either as a paraphrase of the last part of a short story, or a description of a set of pictures. In all these tasks results show that the capacity of making distinctions between the meaning of the connectives develops much more slowly than one would predict on the basis of the first experiments dealing with comprehension. Finally, in a sorting task where the connectives are presented in isolation, even at the age of eleven to twelve years, children's classifications are largely based on physical similarity, and apparently only in a limited way on semantic knowledge of the words. But at the same age level and in the same kind of task, performance with other lexical categories such as concrete nouns or adjectives is guided primarily by semantic criteria.

How do we account for the differences in performance observed with the same connectives in the different tasks? We have already pointed out that in acting out sentences the child may make use of the information provided by the sentences as a whole without processing the connectives, or that he may have only partial knowledge of the connectives which allows him to succeed when contextual information is provided.

According to the first hypothesis, correct performance in the acting out task would be mainly based on information given in the two clauses and the implicit relation between the two events described. Knowledge of the meaning of the connective itself would not be essential. Specific experiments on this issue[1] have shown that children make many more errors in an acting-out task with causal sentences involving relations of low pragmatic plausibility, in which the causal relation is rather arbitrary, for example, "The boy

drinks the milk because the dog follows the cat" vs. "The glass broke because
it fell off the table." This result indicates that the degree of knowledge
of the meaning of connectives which is necessary to correctly perform a comprehension task--and probably to correctly comprehend a sentence--is a
function of the semantic information given in the sentence and the pragmatic
relations reflected in the sentence. Most of the sentences used in the experiments summarized at the beginning of the paper had at least some pragmatic
plausibility and were semantically rather well integrated. So we cannot exclude the possibility that knowledge of the connectives required for comprehension of the sentences did not need to be very elaborate.

According to the second hypothesis, the child would have only "incomplete"
knowledge of the meaning of the connectives. This knowledge would suffice
to allocate the correct meaning to the word when contextual information is
provided and/or an appropriate syntactic frame is given, but would not guaran-
tee success in tasks where words are used in isolation, or when judgments
are called for about the appropriateness of a word in a given context. Even
if the child's syntactic knowledge is well developed and he has full understanding of the meaning of the connectives, he might still not be able to
correctly perform tasks requiring judgments of appropriateness or of meaning
similarity. This kind of performance might well require a more abstract
level of knowledge, of a more metalinguistic character. The task of choosing
between two sentences which are identical except for a critical word, for
example, conceivably requires the capacity of "extracting" the critical words
out of the two sentences and comparing their meaning. This additional operation may account for the critical difference in performance. Notice that
at the age where performance differences among the tasks were found, children
are already capable of constructing sentences by filling in "slots," making
word substitutions, etc. as is done in grammar exercises at school. But the
tasks used in our Experiment 1 through 4 perhaps require more. First, the
child has to identify the function of the critical word in the sentence.
Second, he should be capable of "simulating" a substitution of the word with
another word of the same class. It might be that this more abstract operation
is more difficult, especially if this operation is not carried out with items
such as concrete nouns, which have an easy referent, but with function words
such as connectives. Finally, the sorting experiment requires knowledge of
the meaning of the connectives completely out of context. While such knowledge may be accessible at a relatively early age for words having a "constant

value" such as nouns, it may develop rather slowly for the more "relational" types of words.

Notice that even if we could explain the correct performance in the acting-out task at the age of eight as entirely due to the use of contextual information, we still have to account for the difference in performance between Experiments 1 through 4 and Experiment 5. While many fourth graders and sixth graders seem capable of differentiating between the meaning of the connectives embedded in sentences when they are asked for judgments of appropriateness, most six graders do not yet seem able to "classify" connectives on the basis of semantic distinctions (as they are capable of doing for concrete nouns). The most plausible conclusion here seems to be that the two types of task require access to different levels of semantic knowledge, that these develop at different rates, and that acquisition is a slower process than one could predict on the basis of results such as those reported in the first part of the paper.

The process by which the child acquires meaning distinctions among "difficult" words such as the connectives we have been studying seems to be a slow one. Knowledge of these words necessary for comprehension and production of spoken or written language seems to be available much earlier than knowledge necessary to use them in tasks of a more metalinguistic character. When we talk about the acquisition of word meaning we should be very aware of this. Using a word in a sentence differs from making judgments about it as in our tasks and, in general, from deciding whether it has the same meaning as some other word or not.

*ACKNOWLEDGEMENTS*

Parts of the work reported here have been supported by Z.W.O. Grant No. 35/24 from the Dutch Government to the author. Preparation of this paper has been wholly supported by the Max-Planck-Gesellschaft, Projektgruppe für Psycholinguistik, Nijmegen. Experiments 1 and 3 were carried out while the author was at the University of Padova. Grateful acknowledgement is extended to the Institute of Pedagogy of Padova University for its support in obtaining access to the schools where this research was conducted. Experiments 2, 4 and 5 (with the exception of the pilot study) have been carried out by Mrs. G.A. Damming-Nijrees, under the author's proposal and supervision, as part of her work towards an undergraduate thesis. For the execution of these experiments, she should receive full credit. The work will be reported in greater detail in G.A. Damming-Nijrees, "Verwerving van de betekening van onderschikkende voegwoorden," Undergraduate thesis, Psychologische Funktie-leer, University of Leiden. Finally, the author is indebted to Robert Jarvella and Anne Sinclair for valuable help in improving the structure and the read-ability of the paper itself.

*FOOTNOTE*

1. The research referred to has been carried out by J. Joustra at the University of Leiden, and will be reported elsewhere.

*REFERENCES*

Anglin, J.M. *The growth of word meaning*. Cambridge, Mass.: M.I.T. Press, 1970.

Chapman, J. The acquisition of semantic structures by young children. Doctoral dissertation, University of Birmingham, 1976.

Flores d'Arcais, G.B. The acquisition of the subordinating constructions in child language. In: R.N. Campbell & P.T. Smith (eds.), *Language development and mother-child interaction*. New York: Plenum Press, 1978(a).

Flores d'Arcais, G.B. The comprehension of causal sentences in children. In: G. Drachman (ed.), *Salzburger Beiträge zur Linguistik*. Salzburg, 1978(b).

Flores d'Arcais, G.B., Joustra, J., & Joustra-De Boer, M.L. Children's comprehension of causal sentences. Unpublished manuscript, University of Leiden, 1976.

Johnson, S.C. Hierarchial clustering schemes. *Psychometrika*, 1967, 32, 241-254.

Kruskal, J.B. Multidimensional scaling by optimizing goodness of fit to a non metric hypothesis. *Psychometrika*, 1964, 24, 1-27.

Uit den Boogaart, P.C. *Woordfrequenties*. Utrecht: Oosthoek, Scheltema & Holkema, 1975.

Vygotsky, L.S. *Thought and language*. Cambridge, Mass.: M.I.T. Press, 1962.

# The Metalinguistic Vocabulary of a Speech Community in the Highlands of Irian Jaya (West New Guinea)

Volker Heeschen

Ruhr-Universität Bochum, D-463 Bochum, Fed. Rep. of Germany

Like experiments in psycholinguistics, languages being described for the
first time can serve as testing grounds for general linguistic hypotheses
and theories.[1] A basic assumption of works written within the transformational generative paradigm is that a grammar "is *descriptively adequate* to the
extent that it correctly describes the intrinsic competence of the idealized
native speaker. The structural descriptions assigned to sentences by the
grammar, the distinctions that it makes between well-formed and deviant,
and so on, must, for descriptive adequacy, correspond to the linguistic intuition of the native speaker" (Chomsky, 1965, p. 24). The primary facts, then,
upon which these grammars are based, are the native speaker's tacit knowledge
of his language, knowledge which can be made explicit through judgments of
grammaticality and acceptability of well-formed and deviant sentences. This
assumption is evident in some psycholinguistic work. To quote only one example (Brown, Cazden, & Bellugi, 1971, p. 383), research on the development of
grammar in children is "directed at two general questions: What does the
child know of the structure of English at successive points in his development? By what processes does he acquire his knowledge?" According to Brown,
et al., the possibility of answering these questions is intimately linked to
acceptance of the generative model: "The most demanding form in which to pose
the question of the child's knowledge of structure at any time is to ask for
a generative grammar that represents his knowledge" (p. 383).

As is well-known, definitions of the terms *knowledge, tacit knowledge,*

156

and *intuition* are inextricably linked with the competence-performance dichotomy, the hypothesis of innateness of language and language structure, the
problem of psychological reality of linguistic structures, and the assumption
of a language acquisition device (see Hörmann, 1976, pp. 35-59, and Levelt,
1975). In the light of studies not done in the transformational framework,
I believe that undefined cover-terms (*competence, knowledge,* and *tacit knowledge*) which imply a priori principles of language structure and language
acquisition may simply turn out to be principles a posteriori, which presuppose the development of certain cognitive, pragmatic and communicative
abilities (Bates, 1976; Greenfield & Smith, 1976; Halliday, 1975; Moerk &
Wong, 1976).

On the other hand, linguistic field workers have dispensed with some of
the basic assumptions of Chomsky's theory.[2] Though the body of work that
uses one of the various generative models as a descriptive device is still
growing, nobody seems willing to rely on the native speaker's intuitions or
knowledge. Sommer (1972, p. 8) claims that a Fillmore-type grammar explains
the native speaker's intuition more adequately than the tagmemic approach or
the *Aspects* model. However, like all recent field workers, he bases his
grammar on a traditional analysis of a given corpus. Mirikitani (1972)
relies on native speakers' judgments only when complicated problems cannot
be solved by formal, grammatical, or semantic criteria. Driever (1976),
aiming to distinguish acceptable from unacceptable forms, bases her description on data from traditional informant sessions where subjects judge utterances. These judgments supposedly reflect knowledge common to different
native speakers. However, because judgments vary, she is forced to interro-
gate a large number of subjects, and then to correlate variations in judgment
with the speakers' social status. The study of linguistic competence thus
turns into a report on sociolinguistically determined performances.

On the whole, field workers test only the applicability of a linguistic
theory. The selection of a theoretical framework is based on the understanding that a model will permit "some interesting generalizations" and that it
is "adequate for describing the material at hand" (Williams, 1974, p. 2).
Having run through more than 50 grammars written in the last decade I believe
that the applicability of such theoretical frameworks cannot really be tested.
Adequate presentation of the data is largely underdetermined by the theory,
and theoretical frameworks can be applied a priori. Some of the best grammars
I have read defend eclectic positions (Bromley, 1972; Foster, 1969; Osborne,

1974). "Theories" are, so to speak, machines or tools which serve to help manipulate and arrange the language data. Results emerge if one has a *special* question in mind and a *special* problem to solve (cf. Garvin, 1967). The fact that field workers are generally unwilling to believe that their informants possess a somewhat mysterious linguistic competence may be due to their careful elicitation procedures. Hall's summary of experience in the field (1968, p. 73) concurs with Labov's recent reflections (1975) on the notions of intuitions and knowledge: "Naive speakers' awareness of what either they or others say is at best shadowy; when they attempt to introspect, they are likely to dredge up, not more accurate observations, but more confused and inaccurate impressions" (see also Wong, 1975, pp. 43-45). Some linguistic field manuals (Healy, 1964, 1975; Samarin, 1967) do not even mention the possibility of working with the introspections of informants.

These considerations--the vague status of the concepts of competence and (tacit) knowledge in psycholinguistics as well as the inapplicability of these concepts in linguistic field work--do not, however, prevent subjects from being able "to report on their grammar" (Labov, 1975, p. 36). The native speaker's developing interest in the structure of his language and his ability to reflect upon it are legitimate objects of study in their own right, despite these considerations. Reflections on language, seen as a "special kind of language performance" (Cazden, 1976, p. 603), are a "subcase of the ability to extrapolate on experiential data or to be inductive" (Ganz, 1971, p. 6).[3] In contrast to the capacity of introspection and/or linguistic competence (the assumption of which is saturated with philosophical presuppositions), metalinguistic awareness seen as a special skill lends itself to empirical research and remains within the realm of psychological study. Ample though scattered evidence is available as to the existence of the component skills supporting language use during the child's acquisition of language, such as "the ability to recognize, remember, imitate, discriminate, and the like" (Ganz, 1971, p. 6). In some sense all of these activities require the ability to make judgments on language, its structure and function:

"Das Kind entwickelt schon im Alter von ein bis drei Jahren--genauso wie ein Linguist--eine eigene Vorstellung davon, wie seine Sprache syntaktisch und semantisch gestaltet ist und welche pragmatischen Anwendungsmöglichkeiten sie bietet. Das Kind reflektiert also seine eigene Sprachstruktur" (Levelt in: Max-Planck-Gesellschaft Presseinformation, 1977, no. 8, p. 3).

Studying awareness or self-reflectiveness may prove to be an inspiring method for investigating the child's inductive reasoning and ability to generalize. It may also provide information concerning the degree of accessibility of various levels and hierarchies of language. Even knowing what remains inaccessible would be a desirable and interesting result.

To my knowledge, the question of self-reflectiveness has not yet been raised explicitly in anthropological linguistics. The present paper is in line with Stross' (1974) work on Tenejapa Tzeltal metalinguistics and with studies concerning verbs of saying and speaking in Papuan languages (Deibler, 1971; Pawley, 1969, pp. 27-35). A general rationale is that the native speaker's reflection on and conception of language may prove to be useful for the eliciting linguist.

In this article I present information concerning the adult's awareness of language and speech in a non-western, non-acculturated society. Firstly, I will describe the metalinguistic vocabulary of the community; secondly, I will evaluate the findings in light of other work in anthropological linguistics. The studies I will refer to were written mainly with a view to anthropological, sociolinguistic, or even purely linguistic ends. However, some evidence for metalinguistic awareness can be extracted from the data. I believe that the range of metalinguistic awareness brought to light by studies concerning vocabulary, "linguistic belief systems" (Kernan in Slobin, 1967, pp. 50-51), and "folk linguistics" (Hoeningswald, 1971) can only contribute modestly to the questions discussed so far. However, it is plausible that cross-cultural comparisons and examination of different linguistic repertoires in this regard may yield potentially interesting results.

*THE METALINGUISTIC VOCABULARY OF THE EIPO SPEECH COMMUNITY*

The Eipo live in the Eipomek valley in the Eastern Highlands of Irian Jaya (West New Guinea) at approximately 140° eastern longitude and 4°28' southern latitude. In 1974, when the first members arrived from the interdisciplinary research project "Man, Culture, and Environment in the Central Highlands of Irian Jaya," the Eipo has previously had slight contact with the Frenchman Pierre Gaisseaux, an Indonesian expedition, and some missionaries. The traditional culture of the Eipo has not been affected in any way. They are neolithic horticulturists like all the groups in the mountain regions of New Guinea. Children's upbringing is characterized by a long period of breast feeding (up to three years) after which they develop by actively participating in community life. The Eipo cannot write; non-verbal communication plays an important role in their society of constant face-to-face interaction. Non-verbal symbolism is used on certain occasions, e.g., if someone wants to refute an accusation, he will fasten a bunch of banana leaves to his hut (Schiefenhövel, personal communication). I lived in a native hut in the village of Dingerkon. During the first months, formal informant sessions and work using tape recordings were impossible: data were collected by participant observation. The approach was monolingual. Until now nothing has been published on the grammar of the Eipo language (nor the corresponding language family as a whole). It should be noted that the vocabulary presented here was not collected for the purpose of writing an article on metalinguistics. For a more detailed account of the research project and of the culture of the Eipo, see Koch (1977) and Schiefenhövel (1976). Defining metalinguistic terms as a subclass is not easy, in spite of the existence of informal definitions, as for example that the words and the meaning included are used for thinking about language structure and use, and for commenting on it (cf. Read, this volume); or that the words and expressions "all have in common the feature of referencing some aspect of the communication system--whether an utterance previously spoken, element in the code, or the major patterns of the grammar of interpersonal relations in the society" (Sanchez, 1975, p. 163). There are no clearcut differences, but rather clines between the words referring to linguistic acts. We can, however, distinguish the following categories:

1) verbs of knowing and inner experiences, which can be more or less closely related to spoken utterances and judgments;

2) words for physiological states and activities like *siiandika* (or *siingangdik*)'to yawn' and *iisin* 'to breath, to gasp' (while speaking), which can either accompany verbal utterances, or which function as stand-ins for the verbs of saying;

3) words for non-verbal communicative acts like *sii leb-*[4] 'to smile' and *asing iikin-* 'to wink;'

4) use of onomatopoeia such as *hanghangana* 'to shout ha, ha' (during the dance), *eeana* 'to shout eh, eh' (during the dance)[5] and *maning waiwaiwai-bil* 'the grass is rustling;'

5) sentence and discourse final particles like question markers, tag question markers, quotatives, and particles emphasizing the truth of an utterance or expressing the attitude of the speaker to what

has been mentioned;

6)  inflectional verbal categories of mood, which range on a low level
    of self-reflectiveness of what is appropriate during a speech event.

In the list of metalinguistic vocabulary presented I have excluded categories (2), (3), and (4). Most of the verbs in (2) and (4), to which we can
assign the feature [-linguistic] according to Ross (1970), never take the
affixes of aspect, tense and person which co-occur with the other verbs.
Apparently it is not much use to be able to refer to somebody's yawning or
sneezing in the past or in the future, and reference to person is clear from
the context. In this particular case the lexicologist uses a formal criterion. Whenever an expression like *sii leb-* 'to smile' is a composite form,
both parts (*sii* 'tooth, name,' *leb-* 'to uncover, speak') are included in the
list. The classification of the vocabulary according to certain topics is
very tentative. Some words have too wide a meaning to be subsumed under
one heading. Moreover, the non-alphabetical listing has some disadvantages.
For example, the fact that *yupe* (see no. 12) is one of the few words productive in the formation of new expressions is obscured, and the *derivational
motivatedness* of some words cannot be immediately evaluated (see nos. 21,
40, 92). Within each subsection, the vocabulary is alphabetically arranged.
The examples given serve as sample proofs and as illustrations of the fact
that it is not the vocabulary itself, but its actual use, that creates metalinguistic statements.

*THE VOCABULARY*

*The Metalinguistic Vocabulary Proper. I: Speech, Name, and Meaning*

1.  *diib yupe* 'true, constant language.' In opposition to tabooed,
    special, or obsolete words and expressions in songs, myths, and
    dancing songs *diib yupe* refers to the language of all communication.
    Cf. *diit sii* (no. 46) and *diib sii* (no. 61).
2.  *kekene* 'something carved, meaning.' *malye diit kekene gum* 'a bad
    song, it has no meaning' (it has not been elaborated). *kekene* is
    the name, too, for the big, carved war shields in the neighboring
    valley of Tanime, which are the most elaborate artifacts the people
    in the area possess. Cf. *web-* (no. 20).
3.  *lik yupe* is the name the Eipo usually give to their dialect. *na lik*
    'I don't want, I'm afraid.' In Eipomek three dialects are spoken,
    the Eipo dialect being the easternmost of a western dialect group,

the other dialects belonging to the dialect chains in the east and in the south. Traditionally the Eipo have been making war against the western neighbors. Now instead of *lik yupe, eipe yupe* is currently used, but I believe that this name has been introduced by white people. Cf. *ware yupe* (no. 10) and *una yupe* (no. 9).

4. *merkem* 'clever, skillful, correct,' e.g., in fighting or speaking: *eipe yupe merkem* 'he is clever at speaking the Eipo language.' Cf. *duukdukal-* (no. 68) and *teleb yupe* (no. 8).

5. *nonge* 'body, trunk, the edible part of a fruit, principal part. kernel of a nut, meaning.' *bolkurenang nonge wik* 'the white people have a big body,' *bama donoklyam nonge diblyam* 'throw away the skin, eat the fruit.' When opposed to *sii* 'name,' *nonge* stresses the meaning of a word. E.g., when comparing Eipo *binnab* and Tani *binkayeb* which both mean 'we will be going, let us go,' the Eipo can say *nonge neka sii elebeleb* 'the meaning is the same, the name is different.' Cf. *sii* (no. 6).

6. *sii* 'tooth, mouth, name.' Most currently used word in the frame "tell me the name of ___." As opposed to *nonge* (no. 5), it stands for the mere name of a thing, though no real opposition between the concept and the acoustic form is established. *el sii betinye nonge neka* 'it has two names, but the meaning is the same.'

7. *siknangsusuk* 'sisters and brothers of the same sex, relatives, synonyms.' Cf. *weicape* (no. 11).

8. *teleb yupe* 'good or kind words.' *bilebuk teleb yupe lenmum* 'when he left, you said kind words (to him).' Can be used, too, for stating an utterance to be grammatically correct. Cf. *kin-* (no. 95) and *merkem* (no. 4).

9. *una yupe* 'the dialect of Larye.' *una Larye* 'what?' (*yate* Eipo 'what?'). Cf. *lik yupe* (no. 3) and *ware yupe* (no. 10).

10. *ware yupe* 'the dialect of Tanime.' *ware Tani* 'what?'. Cf. *lik yupe* (no. 3) and *una yupe* (no. 9).

11. *wikyape* or *weicape* 'younger sisters and brothers, synonyms.' *siknangsusuk* (no. 7) is used, when there are two synonyms, *wikyape,* when there are more than two words related in meaning.

12. *yupe* 'sound, utterance, word, speech, language.' *yal ma el yupe fuk fuk fuk alamle* 'the *yal* bird is usually uttering *fuk, fuk, fuk,'* *yupe teleb lebmalam* 'you are using a correct word,' *yupe teleb leb-*

*maseak* 'they are speaking good words/making a good speech to us,'
*nun yupe* ____ *winyalamak* 'in our language one usually says ____.'
Thus *yupe* covers all kinds of articulated sounds and its meaning
is extended to sounds; it is opposed, however, (1) to *diite* 'song'
and *mot* 'dancing'-- *diit gum yup-ak lenmasil* 'no song, language
only he is speaking to us;' (2) to the non-intelligible babbling,
scolding, and crying of somebody, which is referred to as *yupe el
kwit* 'the younger brother of language.'

## *The Metalinguistic Vocabulary Proper. II: Verbs of Saying*

13. *ab-* 'to make, do, say.' *male biisiik alamak* 'they bend the bow'
    (lit. they make the direction for the arrow), *mem abmal* 'he says
    no,' *ur abmal* 'he says yes,' *war abnamlul* 'he might be speaking the
    *war* dialect' (of Tanime).

14. *leb-* The primitive meaning is 'to uncover, to lay bare.' This is
    clearly transparent in *ame len-* 'to peel taro' (with stem change
    for repeated action) and in *sii leb-* 'to smile' (lit. to uncover
    the teeth). Thence: *yupe leb-* 'to speak' (lit. to uncover speech)
    and *diite len-* 'to sing a song.' Finally, *leb-* alone has the
    meaning of 'to speak': *kanye tenebuka lenmarik* 'having reflected
    the two of them said,' *teleb lenmalam* 'you are speaking correctly,'
    *liiman lenmal* 'it is thundering,' *lake leb-* 'to speak openly,'
    *teleb lebne* 'I have spoken well' (closing formula of a story).
    In the Tani dialect, where other words for talking are used, the
    root *leb-* yields the expression for Eipo *asing ketenang* (no. 101)
    'the wise men' (lit. those with opened eyes): *asoru lenning*.

15. *lebareb-* 'to give to speak, repeat.'

16. *nuk-* 'to confer, account, allow, tell.'

17. *nukun-* 'to be telling.'

18. *ngel-* 'to shout.' Specially used for the yelling of the women
    during the dances and of the cultural heroes, who, according to
    legend, left food on the ground and called for Man to rise up out
    of the earth.

19. *siisiib-* 'to name.' Most currently used in the frame "tell me the
    name of ____."

20. *web-* 'to cut, carve.' Thence: *maning webmal* 'he is stripping off
    the grass and rooting it up,' *niinye webuk* 'the man was buried' (the

Eipo bury their dead on a scaffold erected in the top of a tree),
*male wenmal* 'he is carving arrows,' *basam-ak kwaning webmal* 'he is
throwing/preparing sweet potatoes for the pigs,' *mote webmal* 'he is
creating/singing the text during a dance,' *Larye-nang una yupe web-namyak* 'the people of Larye might be speaking the *una* dialect.'

21. *winyab-* 'to do like this, say.' As opposed to *leb-* (no. 14) and
*ab-* (no. 13) most currently used word for 'to speak' (1) in the
frame, ____ *winyalamak* 'they usually say ____' referring to how
speakers of other dialects or idiolects name a thing or pronounce
a word; (2) as a quotation marker in texts: *aleng a-bomsin winyablum*
'I should have put the net bags there, you said.' *winyab-* is a
composite form of *ab-* (no. 13) and *wine* (no. 50).

22. *ura-* 'to say *ura*, repeat the important words or sentences of a previous speaker, chatter, do nothing, be lazy.' To make the wide
range of meaning clear the reader has to imagine an afternoon scene
in the men's house, when the men are chatting, sleeping from time
to time, and eating: a first speaker will tell something, a second
and third giving remarks and comments, the fourth speaker singling
out, and focusing upon, the important words and sentences by repeating them word for word. This scene finds the men in a relaxed
atmosphere, when they have done their day's work. An *urona* is a
person in a begging position, usually a boy or a young man living
for some time in another village, where he is supposed to make
friends.[6]

23. *uruk-* 'to tell.'

24. *yupe areb-* 'to speak, to promise' (lit. to give words).

25. *yupe leb-* 'to speak.' Cf. *leb-* (no. 14).

26. *yupe siidiikin-* 'to tell' (lit. to enumerate words).

27. *yupe tap-* 'to speak' (lit. to tie/bind words together).
*Fol yupe gum tamle* 'Fol has not yet spoken one word.' *niinye yupe
tanmak* 'the men are conversing/talking.'

See also *lube arjuk-* (no. 114) and *yupe wab-* (nos. 37, 86).

*Language as Connected with, and Opposed to, Noise, Scolding and Unintelligible
Speech*

28. *asim yupe* 'language of scolding.'

29. *bokomana* 'din of voices.'

30. *engeb-* 'to cry, weep.'

31. *lelebsan yupe* 'language of scolding.' Grumbling repeatedly *sanib* 'tabooed name of the cassowary' or *basam kalye* 'sacred heart.' Explicitly judged to be *yupe el kwit* 'the younger brother of language.'

32. *meyal* 'noise, clamor.'

33. *metek yupe* 'non-intelligible utterance' (lit. small language).

34. *motmorana* 'grumbling, whispering.'

35. *teik-* 'to scold, call names.'

36. *tekin-* 'to scold, call names.'

37. *wab-* 'to scold.'

38. *wane/wanye yupe* 'insult, insulting speech.'

39. *yupe kwalel-* 'to speak simultaneously.'

See also *ngel-* (no. 18), *yupe gum* (no. 83), and *yupe wab-* (no. 86).

*Language as Connected with, and Opposed to, Music, Song, and Dancing*

40. *bingkon winyabren-* 'to play the Jews'-harp.' Cf. *winyabren-* (no. 92).

41. *diiliib-* 'to sing a song, to repeat a person's words, copy, to learn a language.' As opposed to *leb-* (see no. 14) in *diite lelamle kil* 'the woman who has composed the song,' *diiliib-* stresses the fact that one is only copying the song.

42. *diite* 'song.'

43. *diit amwe* 'the end of a song.'

44. *diit kiisok* 'the beginning of a song.'

45. *diit len-* 'to sing/compose a song.' Cf. *diiliib-* (no. 41).

46. *diit sii* 'name or word, which occurs only in songs.' The words thus referred to are no longer understood by the younger generation. In my opinion these words name persons, geographical places, and metaphors. They have the phonological shape of the idiolects still spoken by the older people. Their songs, however, are partly taboo. The songs mention too many things connected with people already dead. *kam-nang-tam diit teleb* 'the songs of those on the living side are good,' the Eipo say. *diit sii* also refers perhaps to the technique of double meaning excessively used in songs; e.g., *kambulkau kwanikdobre* means literally 'crisping (the leaves) of the *kambulkau* tree,' but as *kambulkau* stands as pars pro toto for

'love song,' the second meaning of 'singing a love song' has either intentionally or unintentionally developed. *diit sii* are *kwanikdob-* (no. 48) and *yukdob-* (no. 52). Cf. *diib yupe* (no. 1) and *mem sii/ mem yupe* (no. 122).

47. *fuana* 'incantation.' In this speech event, too, many obsolete words are used, but they are never said to be a *diit sii* (no. 46). The *fuana* is performed by the *asing ketenang* (no. 101).

48. *kwanikdob-* 'to sing.' Metaphor for singing only used in songs, its primitive meaning being 'to cause something to be crisped or curled.' Cf. *diit sii* (no. 46).

49. *mot* 'dance.' Refers more to the whole setting of a dancing feast and to the movements of the dancers rather than to the verbal sequences, though used in a restricted sense, it refers to that sequence, where the leader of the dancing men is singing. A good *winye* (no. 51) knows the texts very well. Cf. *web-* (no. 20) and *wine* (no. 50).

50. *wine* 'kind of nut, the kernel (edible part) of this nut, words and music of a dance, manner, in this way, thus.' *wine karuakmal* 'he is shelling/cracking nuts,' *mot wine* 'the sung text of the opening sequence of a dance,' *eipo wine bukmalam* 'you are sitting in the way of the Eipo,' *wine alamle* 'he usually does like this.' Cf. *mot* (no. 49) *nonge* (no. 5), and *winyab-* (no. 21).

51. *winye* 'the leader of a dance, the performer of a text sung during the dance.'

52. *yukdob-* 'to sing.' Only overheard in songs, the literal meaning being 'to make something different or beautiful.' Cf. *diit sii* (no. 46).

See also *ngel-* (no. 18) and *web-* (no. 20).

*Topics, Taxonomies, and Genres*

53. *aleng yupe* 'the terminology connected with making net bags.' This type of formation refers to the terminologies connected with certain activities. Other examples: *mot sena yupe* 'the terminology of dancing and of the accessories for dancing,' *yokankan yupe* 'the language of hunting,' *wa yupe* 'language of gardening.' The enumeration of these terminologies is a rather standardized way of speaking. If one asks an Eipo for a story, he will tell some of these termi-

nologies. For illustrating other dialects, the Eipo will try to give one of these terminologies in the other dialect.

54. *fuuruume sii* 'generic name or name of origin of persons, things, animals, and plants.' The difference between the *sii diiba* (no. 61) and the *fuuruume sii* is, e.g., apparent in the naming of insects. Their *sii diiba*, their 'true name,' signifies characteristic sounds they utter or the specific ways they move, their *fuuruume sii*, among other ways of classifying, can indicate the fruit they live on. Thus the *fuuruume sii* of a class of insects is *am dina* 'taro food.' In everyday language, to my knowledge, the *sii diiba* is used. Sometimes the *fuuruume sii* seems to be an old fashioned, obsolete name or word. Even daily activities have such doubles. Whether this indicates a period of dialect mixing some generations ago is a matter of current research. For this question see especially *kwemdina sii* and *mem sii* (no. 57 and 122). Cf. also *diib yupe* (no. 1) and *diit sii* (no. 46).

55. *kulub kulub diit* 'love song.'

56. *kuuwul sii* 'nickname.' Name given to a boy before initiation. The expression yields the idiom *sii kuuwul gum dobnun* 'they will not envy me' (lit. I won't get any nicknames).

57. *kwemdina sii* 'mythical name.' Clans, persons, villages, mountains, victuals, important things and parts of the body have a *kwemdina sii*, which is identical with the *mem sii*, the 'tabooed name.' These names have the phonological shape of the idiolects spoken by the older people and seem to be cognates of the words of everyday life in the dialects east of Eipomek, cf. *balum* 'mythical name of the penis' with Bi (a dialect spoken some 50 kilometers to the east) *balape* 'penis gourd.' This suggests that some generations ago people of the east migrated from the east and met speakers of the western dialect chain in Eipomek. The cultural predominance of the east (cf. no. 3) is well documented in other fields of research.

58. *kwemdina siisiine* or *kwemdina yupe* 'myth, tale of origin.' These tales consist largely in simply naming the *kwemdina sii* (no. 57).

59. *male diit* 'song of war.'

60. *nukna, nungna* 'account, story.'

61. *sii diiba* or *diib sii* 'proper or permanent name of a person, of things, animals, and plants.' Cf. *fuuruume sii* (no. 54).

62. *siibane* 'story.'

63. *siidiikne* or *siisiine* 'story, enumeration of names.'

64. *wa diit* 'song about gardening or sung in the gardens.'

65. *winyana* 'story, tale.'

66. *yupe siidiikna* 'tale, enumeration of names.'

See also *diite* (no. 42) *mem sii* and *mem yupe* (no. 122), *fuana* (no. 47), and *mot* (no. 49).

## Characteristics of Voices, Speakers, and Verbal Behavior

67. *asi lel-* 'to inquire again and again.'

68. *duukduukal* 'to stutter, speak non-fluently.' The verb is most currently used in expressions like *keting duukduukalebuk* 'it's noon' (lit. the sun having hesitated).

69. *morob-* 'to beg, request.' Explicitly opposed to *fetil-*, where begging is done by non-verbal behavior only. Cf. *urona* under *ura-* (no. 22).

70. *morone* 'begging, requesting.' Said of a person who is always asking you to give something to him. It should be mentioned that the Eipo put no blame on begging and requesting, but cf. *morope* (no. 71).

71. *morope* 'beggar.' Only overheard in insults.

72. *na ka yupe* 'Two speakers saying the same thing at the same time' (lit. my friend's words).

73. *nasak yupe* 'impertinent, insolent word/speech.' Especially said of young children's speech.

74. *ngaka* 'person who is continuously asking questions.'

75. *talye yupe* 'words or speech that is irrelevant.' Said also of the utterances of a person clinging to another one and continuously begging.

76. *yukub-* 'to call (somebody).'

77. *yupe aruk-* 'to speak in a hoarse manner.'

78. *yupe bokib-* 'to be dizzy because of somebody's verbiage.'

79. *yupe dik* 'a failing voice.'

80. *yupe dikle* 'the voice fails.'

81. *yupe dobrob-* 'to call, shout.'

82. *yupe dolon-* 'to shout from far away.'

83. *yupe gum* 'to be silent, to be unable to speak' (lit. words/speech not). *el yupe gum* 'he is silent,' *metek me yupe gum* 'the small

child doesn't speak yet.'

84. *yupe gumnye* 'a taciturn person.'

85. *yupe olbanmal* 'the voice comes through.'

86. *yupe wab-* 'to scold.' Opposed to quarrels, where non-verbal actions are involved.

See also *bokomana* (no. 29), *merkem* (no. 4), *metek yupe* (no. 33), *ngalub-* (no. 99), *ngel-* (no. 18), *teleb yupe* (no. 8), *yupe dolaren-* (no. 100), and *yupe kwalel-* (no. 39).

*Speech Acts*

87. *alabrob-* 'to be vague, evasive, to say no, forbid.' Etymon *alab* 'smooth, slippery.'

88. *ceib-* 'to squeeze, ask.'

89. *coleb-* 'to ask'(lit. to break and uncover).

90. *lebreib-* 'to advise' (lit. to put to speak).

91. *tulub-* 'to peel, strip off, advise.' *mun tulubman* 'I will decide for myself' (lit. I am stripping it off in my belly).

92. *winyabren-* 'to ask, answer.' *an-yak winyabrennamkin* 'I ask you' (lit. I will make you say), *winyabrennilyam* 'give me an answer' (lit. put it into words for me). The general meaning is 'to cause to make a sound,' cf. *winyab-* (no. 21) and *bingkon winyabren-* (no. 40). The definite meaning depends on the interplay of the different aspect-tense infixes and the infixed pronouns.

93. *yupe leleb-* 'to advise, consult' (lit. to accumulate words). Cf. *doa motokwe-ak lelebmal* 'the clouds are accumulating on the mountain.' Cf. also *lelebsan yupe* (no. 31).

94. *yupe sek-* 'to advise' (lit. to saw language). The main meaning of *sek-* is 'to saw, to kindle fire,' this being done by pulling a rope to and fro under a small piece of wood.

See also *mem ab-* and *ur ab-* in no. 13, *asi lel-* (no. 67), *morob-* (no. 69), *ngaka* (no. 74), and *yupe areb-* (no. 24).

*Situations where Verbal Behavior Is or Can Be Involved*

95. *kin-* 'to be inclined to a person, say kind words or greeting formulas.'

96. *kwaran-* 'to be shy.' Said of a child's behavior on meeting a stranger and referring more to non-verbal behavior than to the verbal utterances.

97. *luuna* 'calling and shouting of a group of men.' When a group arrives from the rain forest this is done to indicate that they are friendly or that they belong to the village.

98. *murub–* 'to conclude peace.' In this ceremony non-verbal as well as verbal acts (e.g., incantations) are involved.

99. *ngalub–* 'to be astonished.' Refers also to the verbal acts co-occurring with the psychological state of being astonished; these are interjections or comments on the object or event causing astonishment.

100. *yupe dolaren–* 'to shout, call.' Especially used for the shouting and calling that goes on between two fighting parties.

See also *fuana* (no. 47), *mot* (no. 49), *teleb yupe* (no. 8), and *ura–* (no. 22).


*Verbs of Knowing, Thinking, and the Like*

101. *asing ketenang* 'those with opened eyes, the wise men, sorcerers.' These are supposed to be in touch with the spirits, and so they are experienced in healing and know the incantation formulas well.

102. *bik–* 'to know.' Opposed to *eli–* (no. 105), *bik–* states what is momentarily known or unknown. *kam binmal-tam na gum bikne* 'where the dog has gone, I don't know that.'

103. *bikeib–, biki–* 'to get to know, learn.' *biisiik gum bikne bikeibnilyam* 'I don't know the way, make it known to me,' *eibkin are bikeiangkin* 'I see you, that is, I am getting to know you.'

104. *eib–* 'to see, know, realize, appreciate.' *diin–* or *diibren–* 'to see, look.' Has no figurative or extended meaning.

105. *eli–* 'to know, be acquainted with.' *kwemdina yupe mape kelape gum elinyak* 'the boys and women don't know the myths.' Cf. *bik–* (no. 102).

106. *feteb–, fetereb–* 'to show, explain.' *biisiik feterebnilyam* 'show me the way,' *mot fetebnamkin* 'I will explain the *mot* (cf. no. 49) to you.'

107. *gekeb–* 'to hear, understand.'

108. *gekeblob–* 'to remember' (lit. to hear and evoke).

109. *gekebren–* 'to listen.'

110. *kailiil–* 'to make known, make familiar with.'

111. *kanye* 'shadow, soul, echo, thought.' *kanye bik–* 'to be reasonable,

sensible' (cf. no. 102), *kanye bindobnil* 'I am frightened/terrified'
(lit. the soul has left me), *kanye gekeb-* 'to understand' (cf. no.
107), *kanye teneb-* 'to think, reflect' (cf. no. 116).

112. *kele* 'knowledge.' *na kele* 'I know' (lit. my knowledge), *kele wamsil* 'we don't know' (lit. knowledge is just now lacking us).

113. *kelub-* 'to explain.'

114. *lube arjuk-* 'to make a mistake, be mistaken, be misled.' Said of verbal and non-verbal mistakes and errors.

115. *lube kib-* 'to forget.' *lube* 'oblivion,' *kib-* 'to give.'

116. *teneb-*, *teinib-* 'to reflect, think.' Especially used (1) in the first person singular as a clause and sentence final operator: *toubne teleb tenebman ur-diblyam* 'the meat is good, I think, you can possibly eat it;' (2) in the third singular preterite tense: *tenebuk obsik* 'intentionally they killed him;' and (3) *kanye teneb-* (see no. 111). A somewhat contracted form of the first person singular, the formation of which is unique among the morphological processes of Eipo, yields another utterance final operator: cf. *tinye* in *Dingerkon yanmalam-tinye* 'probably you are coming from Dingerkon.'

117. *walewal* 'ignorance, stupidity.' *el walewal* 'he is ignorant, uninformed' (lit. his ignorance), *walewal ubmanil* 'I don't know it' (lit. ignorance is being to me),' *walewal kib-* 'to forget' (cf. no. 115).

118. *weneb-*, see *feteb-* (no. 106).

## The Appropriateness of Speech and Speaking

119. *dengne yupe* 'secret, hidden, reluctant way of speaking' (lit. the language of giving/presenting). For evaluation see the next to final section of this paper and no. 121.

120. *dongon yupe* 'secret, hidden way of speaking' (lit. the language of putting something aside). Cf. *kuunuse don-* (no. 121).

121. *kuunuse don-* 'to tell lies, to speak ironically, to use understatement' (lit. to give shadow to somebody). This is the way to speak of gifts, of taking and giving. Cf. *dengne yupe* (no. 119), *dongon yupe* (no. 120), and *nen yupe* (no. 123).

122. *mem sii* and *mem yupe* 'tabooed name and tabooed myth.' Cf. *kwemdina sii* (no. 57) and *kwemdina yupe* (no. 58). For evaluation also see

next to final section of the paper.

123. *nen yupe* 'secret, hidden way of speaking.' *nena* 'gift' and *neb-* 'to become overgrown.' Cf. nos. 119-121.

### The Truth Value of Utterances

124. *bamul kol-* 'to suggest, using non-verbal signs, that one has not told the truth.'

125. *lengdar kanye* 'a lie, fiction' (lit. the thought of dream).

126. *telel yupe* 'a lie, untrue or incorrect speech.'

127. *telel yupe leb-* 'to lie, to give an understatement.'

128. *telel yupe lelamak-siiliib* 'a class of lies, understatement, irony.'

See also *diib yupe* (no. 1), *kuunuse don-* (no. 121), and nos. 132 and 137.

### Discourse Organizing Particles

129. *akonum* 'enough, sufficient, that will do.' *akonum* alone, or *akonum lebne* 'I have said enough,' can be used whenever a speaker has finished, or is no longer interested in a song, an account, or some more extensive remarks.

130. *amle* quotative particle. *buknun amle* ''I am going to sit down,' he said just now.' Same stem as *ab-* (no. 13) with stem change for completive meaning.

131. *dareb* quotative particle.

132. *diibro* 'is that true?' Sometimes continuously interposed from the listener's side. The speaker, before going on, can answer *diiba* 'that is true' or *diib yupe lenman* 'I am telling the plain truth.' *diiba* (cf. no. 1) 'true' plus question marker.

133. *dikse gum* 'indeed, you don't say so, really?' Remark thrown in from the listener's side.

134. *-do/-ro* question marker. Cf. *yanmal* 'he is coming' and *yanmaldo* 'is he coming?'

135. *gumdo* alternative question marker and tag question marker. *yanmaldo gumdo* 'is he coming or is he not coming?' *yanmal gumdo* 'he is coming, isn't he?' *gum* 'not' plus *-do/-ro* (no. 134).

136. *siirya* 'enough, boundary.' *siirya lebne* 'I have said enough.' Discourse final formula on the speaker's side. Cf. *akonum* (no. 129).

137. *tekro* 'is that true?'. Same meaning and function as *diibro* (no. 132).

138. *-tok* 'only this.' Verb and verb phrase final particle. *winyabne-*
     *tok* 'I say only this,' *bikamne-tok teleb lebne* 'to my knowledge I
     have spoken well' (lit. I have known only this, I have spoken well).
139. *ura* emphasizing, affirmative, assenting discourse final particle
     from the listener's side. Cf. *ura-* (no. 22).

See also *tinye* in no. 116.


*Loan Meanings*

   140. *keken-* 'to write, to carve.' Cf. *kekene* (no. 2).
   141. *lirwe diin-* 'to read' (lit. to look at the color).


*DISCUSSION OF THE VOCABULARY*

   The vocabulary given above is not complete because of the very nature of
the formational and derivational process of the Eipo language. The subclasses
noun plus *yupe* (no. 12) or *yupe* plus verb are open classes; this is also true
for the pattern modifier plus *sii* (no. 6). The sentence and discourse final
particles set apart (nos. 129-139), the semantic boundaries of these entries
to other fields of meaning given are fluent. The salient feature of this
system, then, is its openness and its combining with other domains of speaking
and semantic structures. This is true for all verbs of saying (nos. 13-27),
and it is even more evident for words and expressions for meaning (cf. nos.
2, 5, 7, 11, 20, 50). The semantic fields that seem to be most effective
in creating metalinguistic terms are those of carving and cutting (nos. 2,
20, 140), and of laying bare the edible part of a fruit (nos. 5, 21, 50, 91).
Longer bits of information, conversation, and tales are conceived as acts
of enumerating names, or tying together or accumulating words (nos. 16, 17,
26, 27, 31, 58, 62, 63, 66, 93). Words of thinking and knowing (nos. 101-118)
are clearly distinct from this kind of imagery. While one could be tempted
to subsume the metalinguistic vocabulary proper under one ideal image, that
of uncovering or chaining words, the prototype of the thinking and knowing
domain is 'to hold or to reflect the soul or a thought' (no. 111). The
development of *leb-* (no. 14) from 'to uncover language' and 'to speak' finds
its parallel in this field: *bik-* (no. 102) means 'to stick,' cf. *male bikmal*
'the arrow is sticking,' thence *kanye bik-* 'to stick/hold the thought,' and
finally, *bik-* alone stands for 'to know, be sensible.' The discourse or-
ganizing particles reflecting the attitudes of speakers (cf. nos. 130, 131,
138, 139) originate in both semantic fields.

The literal meanings given in the wordlist do not result from fanciful, hazardous etymologies, but reflect actual usage. To cite one other example not mentioned in the vocabulary, *tab-* in 'bind together words' (no. 27) has two other main usages in the majority of the cases noted or overheard: (1) 'to tie together the legs of a dog' (to prevent him from escaping from the village), and (2) in *yan tabnil* 'my leg has gone to sleep/the leg is tied to me.' While metalinguistic statements are composed mostly of the vocabulary described here, this can be enlarged considerably by the creation of new metaphors, cf. *diite talebmal* 'he is humming a song' (lit. catching a song), which I overheard only once, or compare *uukwe wik* 'big fire' and *uukwe metek* 'small fire,' meaning either that somebody still has a lot to say on a subject or that he is coming to a deadlock (W. Schiefenhövel, personal communication); I have noted only two of the numerous metaphors for 'to sing' (nos. 48, 52). The metaphors and the technique of double meaning in the songs (see *diit sii*, no 46) are another good reason to assume that the Eipo are consciously creating, and playing with, language.

*The Interrelatedness of Some Areas of the Vocabulary*

I am well aware of the fact that the classification given above is a mixture of (a) folk-linguistic criteria (see nos. 12, 31, 46, 83) and of (b) more or less traditional distinctions like that of the verbs of saying. In reality, some of the major Eipo speech situations, events and acts resist any classification. Still, in the undifferentiated stream of Eipo utterances one can single out:

(1)   the songs (no. 41);
(2)   the dancing requiring a special setting (nos. 49, 50, cf. also 20);
(3)   verbal quarreling (nos. 28, 31, 35-39, see also 12);
(4)   the talking of the men in the men's house (no. 22);
(5)   the incantations (no. 47);
(6)   the shouting of the warriors making war and of the children playing war games (no. 100, cf. also 97-98);
(7)   and finally, the *siidiikne*, or technique of enumerating terminologies and the special way the Eipo tell their stories (nos. 53, 57-58, 62-63, 65-66).

The vocabulary suggests delicacy, where there are only hints of an emerging, more elaborate system. Thus, the number of words for *story* could be misleading; they merely reflect the fact that the corresponding verbs can easily be nominalized and that the verbs themselves have a wide range of meaning, their shift into the metalinguistic vocabulary being due to recently formed metaphors (these sometimes have only a short-lived success). See nos. 16-17, 19, 21-23, 26, 58, 60, 62-63, 65-66. There are no well-defined genres and no stories in our sense of the word. Stories, that is to say, larger bits of information, are created spontaneously on occasion, the only exception being the *kwemdina siidiikne* or *kwemdina yupe* (no. 58) concerning the origin of the clans and the deeds of the cultural heroes and ancestors, which by and large are a simple enumeration of their comings and goings and of their handing over the victuals to man.

Basically, the judgments on the grammaticality of utterances (nos. 4, 8) and on their truth value (nos. 124-128), as well as the speech act types (nos. 87-94), make reference to features of verbal behavior that are more or less closely linked to non-verbal behavior. For example, in real life, *yupe merkem* (cf. no. 4) 'skillful speech/utterance' or 'correct utterance' is opposed to *duukduukal-* (no. 68) 'to stutter,' making no, or only subsidiary, reference to grammaticality; and *teleb yupe* (no. 8) 'good/correct utterance' is primarily a synonym of *kin-* (no. 95) 'to be inclined to a person, say kind words,' and an antonym of words like *wane yupe* (no. 38) 'language of scolding.' The Eipo do not focus on the form of an utterance, but on appropriateness and content. That is, on extra-linguistic elements.

"Telling a lie" is for the Eipo, above all, a way of speaking which is a consequence of the rules for avoiding certain topics. It is speaking ironically and using understatement, especially as concerns gifts and other kinds of objects that one is supposed to share or expects to receive (nos. 121, 124, 126). For a more detailed account see Heeschen, Schiefenhövel and Eibl-Eibesfeldt (forthcoming). Perhaps focus on the truth value of an utterance implies a society where *non*-face-to-face interaction takes place and where the control of truthfulness is not guaranteed by shared knowledge and the common perceptual field of the participants.

I believe that the small number of words for speech act types can be explained in a similar way. The high number of performatives in Western-type speech communities seems to be a direct outcome of, and strongly correlated with, a high degree of institutionalization, social organization, and hierarchization. In the Eipo speech community, warning is done by simple interjections, gestures, or expressions of the pattern *kwaning fatabsulul* ('we could be lacking sweet potatoes') or *iisa siisiibsulul* ('the spirits would be calling us'), where the grammatical category hortative-deliberative tunes the propositional content for illocutionary force. Nobody has the status, right, or influence in the Eipo community to officially warn. And begging (nos. 69, 75), asking (nos. 67, 74, 92), and forbidding (no. 87) are ways of speaking accompanied by non-verbal behavior and paralinguistic features. For example, clinging to a person, looking for, or avoiding, body contact or eye contact, speaking submissively like a child, repeating utterances, being impolite, and so on. Words for speech act types may originate and flower in societies where some individuals have a certain rank or privileged position and can thus dispense with non-verbal behavior and paralinguistic features. Rules of rank and the fixed constellations of speaking partners would in some sense then replace the "grammar" of interpersonal verbal and non-verbal communicative interaction. Thus, in view of the nature of everyday life in Eipo society, some well-known distinctions do not seem to apply.

*Comparative Notes*

To sum up, the vocabulary and judgments recorded seem to be entirely content-bound, and do not report on language structure proper. The only word the use of which is restricted to metalinguistic statements is the cover-term *yupe* (no. 12). The general picture outlined here coincides with findings in other Papuan speech communities, as far as this can be inferred from running through the dictionaries available.[7] Usually, a noun meaning 'sound, utterance, language' plus a specifying verb covers the largest or the most important part of the metalinguistic vocabulary. Only Toaripi, where there is one word for 'language, tongue, speech, dialect,' another for 'talk, conversation,' and a third for 'sound' (Brown, 1968), is strikingly different. But I suspect the system of a non-sophisticated speaker of German, for example, to be not very different from that of the Eipo. The specific structural patterns of the Eipo language set apart, I expect the German will mention one of the fundamental facts of language in a similar way, by opposing *Name* to *Bedeutung* or *Sinn* or *was das Wort meint* (cf. nos. 5, 6). Interestingly enough, the etymologies of many everyday German metalinguistic terms reveal similar or identical primitive meanings and metaphorical processes.

*The Vocabulary and its Usages: Two Kinds of Awareness*

So far, the discussion has described only a quite natural system: Reference to speech is made in the same way as it is made to, e.g., types of arrows (as well as non-verbal behavior). Asking for a name, making judgments about somebody's skill in speaking and noting his verbal behavior are a part of everyday language. The Eipo's behavior does not show that they have insights concerning the levels between the word and discourse patterns, nor does it indicate that the Eipo are capable of making inductions and generalizations on language data. In this sense, the vocabulary and its use discussed so far show no, or a only low degree of, self-reflectiveness and awareness, insofar as this requires abstraction from the situation and comment on features of language structure or use.

Indeed, whenever the Eipo are detached from content and context of situation, and asked to shift their attention from actual verbal and nonverbal behavior to language structure, they are uneasy and generally unsuccessful. For example, having elicited the form for 'I am going' and trying to get the corresponding future forms, I entangled my informants in a vivid discussion on where I would be going and whether I would come back. (I did get a lot of interesting forms in the second and third person singular.) Asking five informants to judge the acceptability of the 64 possible combinations of the infixed pronouns and the affixes of aspect, tense, and mood, I concluded that about 90 percent of the answers were simply wrong. However, the percentage of correct judgments on clause chaining was considerably higher, and judgments on word order were absolutely correct.[8] Some of the infix and affix combinations are possible but not frequently overheard in discourse, while clause chaining and word order permit little or no flexibility. Potential use--that which is structurally and semantically possible in a language--and strict rules which cannot be violated, may very well relate to different levels of awareness and accessibility. These levels do not only depend on phonology, morphology, and syntax, the traditionally labelled levels of linguistics. Whether a language has fixed rules in a particular subset of morphological or syntactic processes, or the speaker is permitted to exercise individual choice and creativity (cf. Gleitman, et al., 1972), is obviously also important. I would like to suggest that awareness and accessibility should be highest where choice and alternatives are available. In this case, however, "correct" judgments occurred least often when there were alternatives available.

It should be mentioned that corrections and repairs concerning the infix combinations in *natural* situations were a cause of constant joy for me as the eliciting linguist. For example, one morning I said *kapal toubnamle* 'the plane will land,' but the Eipo had already heard the plane. Always on the alert for the coming of a plane and eager at offering repairs they answered *kapal toubnamal* 'the plane is going to land,' with a slight indignation at my less than ideal linguistic (and acoustic) competence. What had not been possible before was now done in a minute. Using some time adverbials I disentangled the difference between the two infixes *-nam-* and *-nama-*. My informants were distracted and shifted their attention and interest constantly, but in applying their knowledge of the language in actual performance (cf. Chomsky, 1965, p. 3), their judgments as native speakers proved to be perfectly reliable. The ability "to extrapolate on experiental data" (Ganz, 1971, p. 6) thus first appeared (or was first noticed by me) when speech acts were closely tied to actual behavior. The metalinguistic vocabulary used in this case (no. 21), however, was no longer part of the unified whole of everyday language and behavior, but used in true metalinguistic statements resuming former experiences. Still, in non-acculturated societies there must be more genuine and stronger motives of awareness and self-reflectiveness than the linguist's boring questions. Among these must be language varieties and linguistic belief systems.

## MOTIVES OF AWARENESS
### Dialects and Multilingualism

Three dialects are spoken in Eipomek (nos. 3, 9, 10). As a rule the father and the mother of a child speak different dialects because of the patrilineal exogamic system of marriages (Schiefenhövel, 1976, p. 272). In Dingerkon children live with their mothers up to the age of four and speak their mother's dialect. But as they are also exposed to other dialects at a very early age they acquire a perfect passive understanding of them. The low level of confusion between the dialects remains a mystery, though historically, mixing can be demonstrated (nos. 46, 57). Apparently, the child will keep to the mother's language as long as the spatial setting and constellation of discourse participants do not change (cf. Oksaar, 1972, p. 138). Again, as with the infix combinations described in the previous section, the adult's judgments on other dialects are useless. Sometimes, the translation equivalent in the other dialect is simply not known. On the other hand, a special frame for talking about other dialects has been developed: the *siidikne* allows the Eipo to give a more or less complete list of

words and grammatical forms (nos. 19, 26, 53, 63, 66). They learn to pronounce the phonemes their own dialect is lacking. To "learn a language" is
taken to mean to imitate or copy the words or sentences of a foreign speaker
(no. 41). The pronunciation of speakers from more distant speech communities
is imitated and laughed at, and awareness displayed of certain fashions or
trends in speech. For example, boys in a certain age group told us that
one used to say *siirya arebkin* ('I have given you enough'), but that they
had consciously changed the expression to *siirya tenebne* ('enough I think')
(Schiefenhövel, personal communication). Young people can also describe
the idiosyncracies of older people. Two special terms, *diib yupe* (no. 1)
and *diit sii* (no. 46), are used either for the current and everyday language or for old-fashioned and obsolete utterances. These two terms (to
my knowledge the only ones available) are used frequently, and their use is
decisive. Speakers are often confronted with questions of naming certain
objects, animals, and plants (nos. 54, 57, 61), and with problems arising
from semantic and formal characteristics of the songs, the dancing songs,
and incantations (nos. 20, 31, 46-47, 49, 57, 121-122). As obsolete expressions have different phonological and morphological shapes and as the technique of double meaning is used frequently in songs, I believe (but have
no definite proof) that these terms cover not only semantics and phonology,
but also, to a very limited degree, grammar proper. When placed in another
speech community where cognates occur below the dialect level the Eipo are
quick at making hypotheses about the possible phonological shape of Eipo
words in the other language. Back at home, they will use some of the new
words, and will give an account of some of the idiosyncracies of the other
language to their own people.

In most cases, however, knowledge of the surrounding dialects and languages is limited to standardized lists and formulas. This agrees with the
findings of Fox (1974, pp. 72-73): "A subject Rotinese never seem to tire
of discussing is...dialect differences. The local dialect serves as a point
of reference, and the real knowledge concerning other dialects is limited.
What passes therefore as information on dialect difference, although rarely
incorrect, is highly standardized." The Rotinese use a special genre and
the technique of "speaking in pairs," that is to say, using two words for
one notion. The second member of a pair originates in, and is looked for,
in the other dialects.

It is an interesting but still unanswered question whether the degree

of multilingualism is accompanied by an increase in linguistic statements
and self-reflectiveness. A special case, which could be an ideal testing
ground, are the Vaupés in the central part of the Northwest Amazon (Sorensen,
1972; Jackson, 1974). There, each individual learns to speak up to four
and even more languages:

> "In the course of time, an individual is exposed to at least two or
> three languages that are neither his father's nor his mother's language.
> He comes to understand them and, perhaps, to speak them. I observed
> that as an individual goes through adolescence, he actively and almost
> suddenly learns to speak these additional languages to which he has
> been exposed, and his linguistic repertoire is elaborated. In adult-
> hood he may acquire more languages; as he approaches old age, field
> observation indicates, he will go on to perfect his knowledge of all
> the languages at his disposal" (Sorensen, 1972, pp. 85-86).

Though Sorensen and Jackson wrote their papers with emphasis on sociolinguistic implications, I believe that the following facts speak in favor of
a high degree of self-reflectiveness among the Vaupés:

(1) The Vaupés keep closely related languages strictly apart (Sorensen,
    1972, p. 82).

(2) People who mix languages up are held at fault (Jackson, 1974, pp.
    62-63).

(3) If a Vaupés learns to speak Spanish, he will not model its syntax
    on Tukano, the lingua franca of the area (Sorensen, p. 83).

(4) The Vaupés stress the mutual unintelligibility of their languages
    (Jackson, p. 61).

(5) "The diverse and discrete phonologies of these languages and their
    dialects loom very prominently in the Indian's regard" (Sorensen,
    pp. 88-89).

(6) A Vaupés is aware of how fluent he is in a language he is learning
    (Sorensen, p. 87).

(7) The application of the newly learned languages is carefully moni-
    tored (Sorensen, p. 87).

As with the Rotinese, the awareness of the Vaupés is mediated through the
sociology of speaking: it originates in the efforts of the Indians to maintain and indicate their tribal affiliation and identification (Jackson, 1974,
pp. 59-64).

*Special Languages, Appropriateness of Speaking and Linguistic Belief Systems*
Two other motives of metalinguistic awareness remain in the background

of the Eipo speech community: complicated ways of speaking and linguistic belief systems. The Eipo put severe restrictions on openly naming certain activities, goods, ancestors, and dead relatives (nos. 119-123). They do not speak openly of events of utmost importance to them. The data of a dancing feast, a feast of sharing and giving, is hinted at only faintly. The men never mention their departure to the hunting grounds. The exchange of gifts takes place non-verbally (nos. 119, 120, 123; see also the section on interrelatedness). If somebody speaks of gifts or of the exchange of gifts, he will do so using understatements: *kuunuuse dobnamkil* "he will give shadow to you" (no. 121). In the hunting grounds, where the spirits live, one has to speak in a low voice and avoid certain topics. The hunters use a special language, that is, different sets of words for relevant objects and actions (cf. no. 1); coming back they will not mention their game. Cultural goods, men's houses, mountains, some animals and plants have a *mem sii* (no. 122, see also 57-58), a tabooed name, which can only be used in the men's house. The Eipo are careful not to name things connected with dead persons, who are believed to have a relation to the spirits; thus one listens to the songs of the older people reluctantly and mistrustfully (cf. no. 46). Naming, then, is a dangerous action and, inappropriately done, it arouses the anger of the hosts or the wrath of the spirits. The word is something to be handled carefully. It is compared to the edible part of a fruit and the kernel of a nut, and speaking itself may be the action of uncovering highly esteemed things (nos. 5, 14, 20-21, 50). Even the word *yupe* (no. 12) fits into this picture; the labialization of a final consonant and its voiced release are a nominalizing device, and *yub-* means 'to cook,' which is done by placing food in hot ashes or by enclosing it in leaves and putting it into the fire. Speaking, then, could be compared to the uncovering of properly cooked food. In view of more complicated ways of speaking, three features in the primitive system of the Eipo are interesting: (1) the value attached to the word; (2) the careful "preparation" of the word according to belief; (3) the resulting need to sometimes handle words carefully.

Hale (1971) reports a special language the Australian Walbiri use when they have reached a certain stage in their initiation rituals. The principle for speaking this language is the following: "replace each noun, verb, and pronoun in *ordinary* Walbiri by an *antonym* " (p. 473). Thus, a sentence like "I am sitting on the ground" should be turned into "He is standing in the sky." Simple as it seems, the principle cannot work when taxonomies like

names of animals, or morphemes like the inchoative and causative-transitive
affixes of the verb, are involved. Concerning the taxonomies, the general
principle is to oppose "entities which are immediately dominated by the
same node in a taxonomic tree" (p. 477). The transitives are replaced by
intransitives, but transitivized verbs like *see* (vs. *cut*) which do not actually produce effects upon their objects, are excluded. The Walbiri have to
learn the skill of a rather penetrating semantic analysis. Their intuitions

"are not inconsistent with semantic structures which have been recognized throughout the history of semantic studies--these include not
only such obvious concepts as polarity, synonymy, and antonymy, but
also the more general conception that semantic objects are componential
in nature, that lexical items often share semantic components and belong to well defined domains which exhibit particular types of internal
structure ..." (Hale, 1971, p. 478).

Once more, linguistic skills develop during a special performance, or the
"perceptual plausibility" (Prakasam, 1976, pp. 323-324) of semantic structures originates in a special language game. Another root of this game may
lie in the subjects' pleasure in rule infraction: giving other, or fanciful,
names to objects for which the subject has mastered the correct designations.

The Dogon (Calame-Griaule, 1968) and the Bambara (Zahan, 1963) have
complicated belief systems about human speech, and have created a real
"mythology of the word." Focus is put on the difficulty in producing the
word; most of the organs in the human body are believed to contribute to its
creation. The Dogon focus on the cultivation the word deserves once it is
pronounced, the power of the word in social life, the care with which it
needs to be handled, and the role it plays at crucial periods in an individual's lifetime. Here let me give the example of the natural phonetic analysis
of the Bambara (Zahan, 1963, pp. 51-61). The Bambara clearly are familiar
with the elements that make up the word. To begin with, they make a difference between consonants and vowels. The five "cardinal" vowels "consituent
les sons fondamentaux du langage, dont, en quelque sorte, ils sont l'âme.
On les appelle 'sons emmanchés' car ils peuvent être prononcés sans le secours
d'autres sons" (p. 51). The consonants are 'sons non emmanchés.' The five
vowels are subject to modifications: they can (1) be shortened; (2) have
more opened or more closed qualities; (3) be nasalized; (4) be diphthongized.
"Diphthonguer la voyelle signifie la *marier*" (p. 55). Three tones are
discriminated: "le 'ton de montagne,' *ka kuluma*, le 'ton d'eau,' *ka dyima* et
'le ton de vent,' *ka fyẽma*" (p. 55). The consonants are differentiated into

182

simple ones and clusters. This analysis, however, is not carried out for
linguistic reasons. Each sound symbolizes magic powers that relate to and
influence human life. For example:

> "La signification essentielle de la nasalisation est l'intégration du
> son dans la vie même de celui qui le prononce, tandis qu'à son tour
> le sujet parlant insuffle sa vie dans le son" (Zahan, 1963, p. 55).

The consonants and the clusters, then, can be arranged according to these
powers and the relationship between them. Taking the cluster *mn* as a point
of departure one arranges the consonants in the following ways:

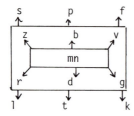 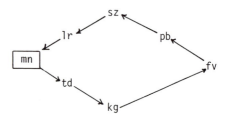

Figure 1. A representation of the
idea of distance from the speaker
in Bambara (from Zahan, 1963).

Figure 2. A representation of
the idea of cyclic movement
around a point of origin in
Bambara (from Zahan, 1963).

The relationships among the sounds is not incorrect in light of modern phonological analysis. But at the same time, as Zahan explains, the judgments
of phonological similarity of the consonants are intricately tied to a
fascinating belief system, having to do with ideas of "distance from" the
speaker (*mn*) and of cyclical movement around a point of origin.

*CONCLUSION*

Evidence from anthropological linguistics shows that self-reflectiveness
and the capacity of making judgments on language are mediated through manners
of speaking, that is, through specific performances carried out in specific
speech communities. The component skills of this ability and ways of induction and generalization develop with the specific demands a society makes on
its members. The motives for a native speaker's genuine and natural interest in his language come from the plurality of idiolects, dialects, languages,
word play, and fashions of speaking; it originates in judgments on the appro-

priateness of the uses of words and in language games. Inasmuch as reference to utterances previously spoken, to their truth value, and to characteristics of verbal behavior are necessary for clarification and orientation in discourse, a metalinguistic vocabulary can be found in many (perhaps all) languages. Inasmuch as language varieties, idiolects, and multilingualism are inherent features in the history of speech communities, self-reflectiveness arises, though its extent and importance are largely contingent. Still in need of exploration are the naive speaker's deep rooted belief in the power of words, the different belief systems it brings on and the awareness it entails. The levels at which there is least evidence of natural metalinguistic awareness are inflectional morphology and syntax. However, the findings discussed do not yield a cohesive picture because the material I have worked with was largely disparate.

In one sense, the approaches of the anthropological linguist and developmental psycholinguist are comparable. Both have to handle naive, unsophisticated informants; both have to look for the pragmatic (social and genetic) prerequisites involved in the development of certain skills. However, there are no obvious functional reasons why the naive adult should attend to the structure of language, as the structure itself is only a means for deployment of other activities. However, there do seem to be signs that metalinguistic awareness in children is useful and perhaps necessary. The wider range of functions of child language (Halliday, 1973), the rich imagery of the child, and incomplete level of automatization of speech production as well as gaps between perception and production may well require a system of coordination and attention shift (Bever, 1975; Bruner, 1966; Dingwall, 1975; Fromberg, 1976; Piaget, 1972). But in light of anthropological findings, it would be premature to assume that the child focuses on those aspects of language favored by grammarians, namely morphology and syntax. Grammar proper as a secondary, derivative and higher level of organization comes into being with a multiplicity of uses in content-bound discourse. The metalinguistic vocabulary of a non-acculturated speech community reflects this aspect of language.

*ACKNOWLEDGEMENTS*

This is publication no. 15, Man, Culture, and Environment in the Central Highlands of Irian Jaya, a research project of the Deutsche Forschungsgemeinschaft. I would like to thank the initiators of the research project,

K. Helfrich and G. Koch, and the D.F.G. for its support of the project.
My thanks are extended to the Lembaga Ilmu Pengetahuan Indonesia (Jakarta),
the Universitas Cenderawasih (Abepura, Irian Jaya), The Missionary Aviation
Fellowship, the Unevangelized Fields Mission, and all others who have assisted
me.

*FOOTNOTES*

1. The idea of using the metalinguistic vocabulary in the way reported
here was stimulated by the Max-Planck conference this book is a reflection
of. The linguistic findings summarized would have been impossible without
the constant and stimulating discussions I have had with W. Schiefenhövel;
some of the very important findings concerning the vocabulary of the Eipo
are due to him, e.g. the important complex  connected with the *asing ketenang*
(item no. 101 of the wordlist). The alphabet used and the main phonetic
realizations of the corresponding Eipo phonemes are as follows:

a [a], b [b,p], c [tj], d, e [ɛ,e,ə], f, g, i [e,ɪ,i], ii [i], k [k,x],

l, m, n, ng [ŋ], o [ɔ,o], p [p$_w$], r [ɾ], s, t, u [o,ʋ], uu [y,u], w [ß],

y [j].

2. The following remarks are based on a manuscript by the author entitled
"Linguistische Theorien und Feldforschung."

3. It should be noted that Ganz' formulations are written in another
context.

4. The citation form of the verb is the stem.

5. Phonetically [haŋhaŋ'ʌnʌ] and [ȩ̃ȩ̃ʌnʌ]. The phone [h] and the laryn-
gealized vowels do not make up part of the phonetic realization system.

6. This ability to repeat whole phrases, sentences, and paragraphs
contrasts peculiarly with initial inability to repeat utterances from tape
recordings. Later on, however, some informants learned to do this, and they
even imitated the paralinguistic features of the recordings.

7. Aufenanger, 1953a and 1953b; Brown, 1968; Doble, 1960, Keysser, 1925;
Lang, 1973; Loving, 1975; McElhanon & McElhanon, 1970; cf. also Deibler, 1971
and Pawley, 1969.

8. The testing was done quite informally and spread over some days. In
one way or another all sessions with native speakers are a kind of testing;

accordingly, we may well suppose that a wealth of information on metalinguistic awareness is hidden in the notebooks of linguistic field workers.

*REFERENCES*

Aufenanger, H. *Vokabular und Grammatik der Nonduglsprache in Zentral-Neuguinea*. Posieux/Freiburg: The Anthropos Institut. (Micro-Bibliotheca Anthropos, 5), 1953(a).

Aufenanger, H. *Vokabular und Grammatik der Gende-Sprache*. Posieux/Freiburg: The Anthropos Institut. (Micro-Bibliotheca Anthropos, 1), 1953(b).

Bates, E. *Language and context: The acquisition of pragmatics*. New York: Academic Press, 1976.

Bever, T.G. Psychologically real grammar emerges because of its role in language acquisition. In: D.P. Dato (ed.), *Developmental psycholinguistics: Theory and application*. Washington, D.C.: Georgetown University Press, 1975.

Bromley, H. *The grammar of Lower Grand Valley Dani in discourse perspective*. Doctoral dissertation, Yale University, 1972.

Brown, H.A. *A dictionary of Toaripi with English-Toaripi index*. Vol. 2. Sydney: University of Sydney, 1968.

Brown, R., Cazden, C., & Bellugi-Klima, U. *The child's grammar from I to III*. In: A. Bar-Adon & W.E. Leopold (eds.), *Child language. A book of readings*. Englewood Cliffs, New Jersey: Prentice Hall, 1971.

Bruner, J.S. *Toward a theory of instruction*. Cambridge, Mass.: Harvard University Press, 1966.

Calame-Griaule, G. *Ethnologie et langage: La parole chez les Dogons*. Paris: Gallimard, 1965.

Cazden, C.B. Play with language and metalinguistic awareness: One dimension of language experience. In: J.S. Bruner, A. Jolly & K. Sylva (eds.), *Play: Its role in development and evolution*. Harmondsworth: Penguin, 1976.

Chomsky, N. *Aspects of the theory of syntax*. Cambridge, Mass.: M.I.T. Press, 1965.

Deibler, E. Uses of the verb 'to say' in Gahuku. *Kivung*, 1971, 4, 101-110.

Dingwall, E.O. The species-specificity of speech. In: D.P. Dato (ed.), *Developmental psycholinguistics: Theory and application*. Washington, D.C.: Georgetown University Press, 1975.

Doble, M. *Kapauku-Malayan-Dutch-English dictionary*. The Hague: Nijhoff, 1960.

Driever, D. *Aspects of a case grammar of Mombasa Swahili with special reference to the relationship between informant variation and some sociological features*. Hamburg: Buske, 1976.

Foster, M.L. *The Tarascan language*. Berkeley: University of California Press, 1969.

Fox, J.F. Our ancestors spoke in pairs: Rotinese views of language. In: K. Bauman & T. Sherzer (eds.), *Explorations in the ethnography of speaking*.

London: Cambridge University Press, 1974.

Fromberg, D. Syntax model games and language in early education. *Journal of Psycholinguistic Research*, 1976, 5, 245-260.

Ganz, J.S. *Rules: A systematic study*. The Hague: Mouton, 1971.

Garvin, P.L. American Indian languages. A laboratory for linguistic methodology. *Foundations of Language*, 1967, 3, 257-260.

Gleitman, L.R., Gleitman, H., & Shipley, E.F. The emergence of the child as grammarian. *Cognition*, 1972, 1, 137-164.

Greenfield, P.M., & Smith, J.H. *The structure of communication and early language development*. New York: Academic Press, 1976.

Hale, K. A note on a Walbiri tradition of antonymy. In: D.D. Steinberg & L.A. Jakobovits (eds.), *Semantics: An interdisciplinary reader in philosophy, linguistics and psychology*. Cambridge: Cambridge University Press, 1971.

Hall, R.A. Jr. *An essay on language*. Philadelphia: Chilton Books, 1968.

Halliday, M.A.K. *Explorations in the functions of language*. London: Arnold, 1973.

Halliday, M.A.K. *Learning how to mean: Explorations in the development of language*. London: Arnold, 1975.

Healey, A. *Handling unsophisticated linguistic informants*. Canberra: The Australian National University (Linguistic Circle of Canberra Publications, A 2), 1964.

Healey, A. (ed), *Language learner's field guide*. Ukarumpa: Summer Institute of Linguistics, 1975.

Heeschen, V. *Grammatik der Eipo-Sprache*. (Unpublished manuscript, 1975).

Heeschen, V., Schiefenhövel, W., & Eibl-Eibesfeldt, I. Requesting, giving and taking. The relationship between verbal and non-verbal behavior in the speech community of the Eipo, Irian Jaya (West New Guinea). Forthcoming.

Hoeningswald, H. A proposal for the study of folk-linguistics. In: W. Bright (ed.), *Sociolinguistics. Proceedings of the UCLA sociolinguistics conference, 1964*. The Hague: Mouton, 1971.

Hörmann, H. *Meinen und Verstehen. Grundzüge einer psychologischen Semantik*. Frankfurt/Main: Suhrkamp, 1976.

Jackson, J. Language identity of the Columbian Vaupés Indians. In: R. Bauman & T. Sherzer (eds.), *Explorations in the ethnography of speaking*. London: Cambridge University Press, 1974.

Keysser, Ch. *Wörterbuch der Kate-Sprache gesprochen in Neuguinea*. Berlin: Reimer, 1925.

Koch, G. Die Eipo. Anatomie einer Steinzeitkultur. *Bild der Wissenschaft*, 1977, 14, 44-59.

Labov, W. *What is a linguistic fact?* Lisse: Peter de Ridder Press, 1975.

Lefebvre, V.A. *The structure of awareness: Toward a symbolic language of human reflection*. London: Sage, 1977.

Levelt, W.J.M. *What became of LAD?* Lisse: Peter de Ridder Press, 1975.

Loving, R., & Loving, A. *Awa dictionary*. Canberra: The Australian National University (Pacific Linguistics, C 30), 1975.

McElhanon, K.A., & McElhanon, N.A. *Selepet-English dictionary*. Canberra: The Australian National University (Pacific Linguistics, C 15), 1970.

Mirikitani, L.T. *Kapampangan syntax*. Honolulu: Hawaii University Press (Oceanic Linguistics, Special Publications, 10), 1972.

Moerk, E.L., & Wong, N. Meaningful and structured antecedents of semantics and syntax in language. *Linguistics*, 1972, 172, 25-37.

MPG Presseinformationen. Forschungsberichte und Meldungen aus der Max-Planck-Gesellschaft, 1977, no. 8, May 5, 1977.

Oksaar, E. Sprachliche Interferenzen und die kommunikative Kompetenz. *Indo-Celtica. Gedächtnisschrift für Alf Sommerfeldt*. Munich: Hueber, 1972.

Osborne, C.R. *The Tiwi language*. Canberra: Australian Institute of Aboriginal Studies (Australian Aboriginal Studies, 55, Linguistic Series, 21), 1974.

Pawley, A. Transformational grammar and the native speaker: Some elementary issues. *Kivung*, 1969, 2, 22-36.

Piaget, J. *Psychologie der Intelligenz*. 4th Edition. Olten: Walter, 1972.

Prakasam, V. Perceptual plausibility and a language game. *Anthropological Linguistics*, 1976, 18, 323-327.

Preston, D.R. Linguistic vs. non-linguistic and native speaker vs. non-native speaker. A study in linguistic acceptability. *Biuletyn Fonograficzny*, 1975, 16, 5-18.

Ross, J.R. On declarative sentences. In: R.A. Jacobs & P.S. Rosenbaum (eds.), *Readings in English transformational grammar*. Waltham, Mass.: Ginn, 1972.

Samarin, W.J. *Field linguistics*. New York: Holt, Rinehart & Winston, 1967.

Sanchez, M. Introduction. In: M. Sanchez & B.G. Blount (eds.), *Sociocultural dimensions of language use*. New York: Academic Press, 1975.

Schiefenhövel, W. Die Eipo-Leute des Berglands von Indonesisch-Neuguinea. *Homo*, 1976, 26, 263-275.

Slobin, D. (ed.), *A field manual for cross-cultural study of the acquisition of communicative competence*. Berkeley: University of California, 1967.

Sommer, B.A. *Kunjen syntax. A generative view*. Canberra: Australian Institute of Aboriginal Studies (Australian Aboriginal Studies, 45, Linguistic Series, 19), 1972.

Sorensen, A.P. Jr. Multilingualism in the Northwest Amazon. In: J.B. Pride & J. Holmes (eds.), *Sociolinguistics. Selected readings*. Harmondsworth: Penguin, 1972.

Stross, B. Speaking of speaking: Tenejapa Tzeltal metalinguistics. In: K. Bauman & T. Sherzer (eds.), *Explorations in the ethnography of speaking*. London: Cambridge University Press, 1974.

Williams, M.M. *A grammar of Tuscarora*. Doctoral dissertation, Yale University, 1974.

Wong, I.F.H. Field procedures in generative grammar. *Anthropological Linguistics*, 1975, 17, 43-52.

Zahan, D. *La dialectique du verbe chez les Bambara*. The Hague: Mouton, 1963.

# Part II

# Theoretical Aspects

# Conceptualization and Awareness in Piaget's Theory and Its Relevance to the Child's Conception of Language

H. Sinclair

University of Geneva, F.P.S.E., CH-1200 Geneva, Switzerland

Two books by Piaget were recently published on the problem of "awareness," each of them the result of a year's work at the Centre d'Epistémologie Géné- tique: the first is called *La prise de conscience* (1974a)--literally trans- lated "On becoming aware"--and the second is called *Réussir et comprendre* (1974b). Personally I think there are good reasons to be surprised at the choice of the topic "becoming aware." On the one hand, this title could lead us to think that the work deals with the philosophical problem of "con- ciousness" for which introspection was supposed to provide the basic data. Would Piaget take introspection as a subject of experimental study? On the other hand, all Piaget's work on cognitive development can be seen as the study of the child's progress towards higher levels of conceptualization, and conceptualization at whatever level would seem to imply at least some degree of awareness. So why devote a number of experiments and a theoretical discussion to the specific problem of awareness?

Piaget, in his introduction to *La prise de conscience* discusses the first point, and explains that his work does not deal with introspection, but that "becoming aware" is a mental activity of a special type that inter- acts with other cognitive activity on which it depends and which it can modi- fy in turn. Indeed, the experiments described concern situations where the child is asked to perform an action (usually with a few objects) whose re- sult is clear to him (e.g., to use a sling to get an object into a box, to

propel a ping-pong ball on a table in such a way that it returns to its starting point, etc.). Sometimes the child succeeds immediately without any trial
and error, and the experimenter asks questions ("How did you do it? How did
it work?") inciting the child to verbalize or to show in slow motion how he
succeeded; usually, awareness in this sense lags considerably behind success
in action. Sometimes the child does not immediately obtain the desired result, so he has to stop and think; then his awareness of the difficulty and,
possibly, of a way of overcoming it can be inferred not only from his answers
to questions but also from his actions. Clearly, the experiments do not deal
with introspection as a philosophical problem, but with the child's growing
awareness of his own actions on objects and of the interactions between objects that may result from them (e.g., when one sets up a series of dominoes
in a certain way, a tap on the first one will make all the others fall down
one by one).

Clarification of the first point mentioned, i.e., the specificity of
what is called "becoming aware" in these books as against "conceptualization,"
"interiorization" and "abstraction" in others, appears more difficult. Though
a careful reading of the two works mentioned can provide an answer, another
text (Piaget, in Inhelder & Chipman, 1976) would seem more explicit. Discussing the problem of structures, Piaget says: "The results of cognitive functioning are relatively conscious, but the internal mechanisms are entirely, or
almost entirely, unconscious. For example, the subject knows more or less
what he thinks about a problem or an object; he is relatively sure of his beliefs. But though this is true of the results of his thinking, the subject is
usually unconscious of the structures that guide his thinking. Cognitive
structures are not the conscious content of thinking, but impose one form of
thinking rather than another. These forms depend on the subject's developmental level and derive ultimately from early organic coordinations" (p. 64).

In other words, cognitive structures are not observable, nor even directly inferrable from a person's actions or from his conceptual representations that can be rendered observable through verbalizations, drawings,
graphs, formulas, etc. They are what give direction to thinking, in the
sense that the subject "knows" that for a certain problem he has to reason in
such and such a way, which is "necessary," or that he cannot draw such and
such a conclusion because it is "impossible," but the subject has no (or only
very limited) conscious access to the mechanisms by which such constraints or
possibilities were constructed. Thus, when in many of his works Piaget speaks

about self-regulation, assimilation, and accomodation, he is concerned with the constructive mechanisms that govern the functioning and formation of cognitive structures, mechanisms that cannot be directly inferred from overt behavior, but whose nature can be postulated from biological parallels, and upon which certain ways of behaving throw some light, especially those behaviors that allow the experimenter to observe how a conflict arises and how it is gradually resolved (Inhelder, Sinclair, & Bovet, 1974). When Piaget speaks about awareness, he means the subject's gradual awareness of the how and why of his actions and their results and of the course of his reasoning-- but not of what makes his way of acting or thinking possible, impossible or necessary. Thus the research on "becoming aware" is to be interpreted as "becoming aware" of the how, and eventually the why, of specific actions and of the how, and eventually the why, of certain interactions between objects. And in this sense, research on awareness through the presentation of specific situations that make the subject stop and think, or through questions of the experimenter that incite the subject to reflect on his activity, constitutes a specific domain that allows yet another enlightening view on the main theme of all Piaget's research and theory-building, that of the construction of knowledge. It furthermore illuminates in a particularly clear way the theoretically postulated manner of functioning of the underlying mechanisms, which are deeply rooted in biology. Indeed, from the results of the experiments on awareness, Piaget postulated a strong analogy between the deep-seated mechanisms of the construction of knowledge and the more easily accessible mechanisms of "becoming aware."

Without going into the details of either the experiments or the interpretation of the results, several important points need to be made, before the possible relevance of Piaget's research for the study of children's conceptualization of language can be discussed.

In the first place, action know-how constitutes by itself an important body of knowledge, even when it remains outside the sphere of "awareness." Except at the earliest sensori-motor levels, such know-how is always accompanied by some conceptual representation. All awareness has its source in know-how, even when conscious conceptualization goes beyond the available know-how and directs the subject's actions (including his mental activities) towards new discoveries.

In the second place, awareness does not simply consist of becoming conscious of actions performed so to speak automatically. There may be a considerable difference between success in action and awareness, and this difference is often not only a difference in time. In the experiment with the ping-pong ball (1974a), for example, children from six years of age are perfectly capable of producing the boomerang effect (propelling the ball forward with a reverse spin, so that when it stops sliding forward it *returns* to its point of departure rolling backwards). But even eight-year olds have difficulty in explaining what happens (despite the use of a half-white, half-black ball whose sliding and rotatory movements can be more easily observed). These subjects will still affirm that the ball started off by rotating forward and then reversed its direction and began rolling backwards. The youngest subjects, even those who succeed several times in producing the effect, and despite the fact that the experimenter asks them repeatedly where they put their finger on the ball, and what movement they made, can say no more than "One has to throw hard," or, at a slightly higher level, "I put my fingers on top and pull them back; I push forward and backwards."

In other words, awareness, in the sense of a conscious conceptualization of a complex activity and its results, goes beyond the capacity to coordinate the various actions. In the example given, the subject may have the specific know-how to produce the boomerang effect without knowing that he gives the ball both a forward impulse and a rotation. Though he may already have conceptualized the fact that balls can slide as well as roll (and bounce, fall, etc.), he need not yet know that they can slide forward while rotating, or as the older subjects say, "slide and hop and turn at the same time." Moreover, in order to understand what happens, he also has to know that at a certain point the sliding movement stops, having lost it impetus, whereupon the rotation takes over. Thus, though all know-how beyond the level of sensori-motor activity is accompanied by some conceptualization, the various concepts, all derived from specific action know-how, have to be coordinated into a coherent system; and only when this takes place can the partial conceptualizations become conscious. This kind of becoming aware thus takes place at a higher level than the first conceptual representations, which follow the coordinations of the actions themselves, but, according to Piaget, all these construction processes are similar in a deep sense. It is this similarity in gradual construction (which can take place at very different levels of development) that confers special importance to experiments concerned with awareness as a specific topic, even though in many other experiments the Piagetian dialogue technique also leads the subject to become aware of certain aspects of action or properties of objects.

The experiments described in *La prise de conscience* and in *Réussir et comprendre* provide insight into the way awareness proceeds from very simple beginnings. According to Piaget, in all intentional actions, at levels well below those of the children taking part in the experiments, the acting subject is aware of at least two things: the goal he wants to reach and, subsequent to his action, the result he has obtained (success, partial success or failure). From these modest beginnings awareness proceeds in two different, but complementary directions. Especially when the action fails, but also when the subject is pleasantly surprised by success, or, at the ages where this can be done, when he is asked questions, or when he questions himself, he will construct a conceptual representation of at least some of the features of the actions he has performed and of some of the reactions and properties of the objects he acted upon. For example, one can observe how a one to one-and-a-half year old baby tries to put a small cup into a bigger cup, and fails because his own hand gets into the way. The baby then briefly puts the small cup into his mouth, in order as it were to verify that the cup is indeed something that can be put into something else. He may then put a finger or his hand into the bigger cup, to make sure that it is indeed a container; whereupon he will try again and maybe succeed (Moreno, Rayna, Sinclair, Stambak, & Verba, 1976). Though at this level the baby probably works out his problem only in actions, without awareness either of the precise position of his hand that made for failure or of the position and shape of the objects, this is already an example of how conceptual knowledge comes to be constructed; the child somehow reconsiders both his own actions and the properties and "behavior" of the objects.

At the ages studied in the two works mentioned, awareness in the sense of being able to answer questions about or to show clearly what happened during a complex activity that either succeeded or failed seems to start in the same way, from the most directly observable characteristics of the subject's actions and the object's properties. Gradually the subject penetrates deeper into the how and why of his own actions, their succession and coordination, and at the same time he acquires a deeper knowledge of the objects' properties and of the physical laws that govern its behavior. Thus, in the ping-pong ball experiment, the children are at first only able to say (or to show in slow motion) that they touch the ball, throw or push it, and that the ball moves forward and backwards. Only much later can they explain or show exactly where they put their fingers on the ball, how they move their hand, and how and why the ball slides forward and then rolls backwards. This twofold process of conceptualization, on the one hand of one's own material or mental actions, and on the other hand of the physical properties of outside reality, is, according to Piaget, the basic construction process of both logico-mathematical systems and of knowledge of physics. However closely linked these two different aspects of conceptualization may be, there is nonetheless a lack of symmetry in that reflection on one's own thought processes at the highest level leads the thinking subject beyond the integration of observable regularities in events to inferences about other possible coordinations and their results. Epistemologically speaking, this accounts for the fact that the scientist's "adaptation to concrete experimental data is a function of the abstract character of the conceptual framework that allows him to analyze and even to apprehend his own experiments" (Piaget, 1974a, p. 281).

Awareness is thus possible at many different levels of cognitive development and will be evidenced in a different manner according to the particular problem the subject is trying to solve; but its progress closely resembles that of all cognitive constructions, at the level of action coordination as well as at that of conceptualization. Neither the reflection on one's own thought processes nor the understanding of physical causality can ever be complete; the biological sources of thought remain forever closed to consciousness, and ultimate understanding of physical reality is equally unattainable, if only because the observing and thinking subject is himself part of the physical universe. When Piaget proposes (1974a, p. 227) that conceptualization is to action as action is to its neurological substrate, he adds that for conceptualization one can speak of becoming aware of actions, and thereby reconstructing and enriching them, but that for action there can be no question of the subject becoming aware of the neurological substrate; instead of a *prise de conscience* in the latter case one should speak of a *prise de possession:* the subject gradually appropriates the neurological functioning and biological regulations, unconsciously using them to construct his actions.

Almost all utterances belong to the category of intentional acts. Within the framework just sketched, their aim and results should be conscious. But what can be thought of as aim and results of a speech act? Superficially, this question does not seem so difficult to answer: when a speaker utters

a command, a question, a wish, or even a greeting, his aim is some reaction
of the person addressed and the result is this person's reaction; failure
would be if the addressee reacted to a demand as if it were a greeting, etc.
However, though verbal behavior may indeed aim at some specific reaction on
the part of another person, this is far from generally true. It is not even
true of the verbal acts of very young children: they may vocalize or produce
utterances when they are alone; in the presence of another person they sometimes do not seem to expect any response, and sometimes they seem to be
perfectly content with some gentle monotonous noises of assent. More important, it certainly does not seem as if communication failure in this sense
is the main reason that incites the child to mastery of his mother tongue.
Thus, if in actions of practical intelligence awareness of aim and result
leads the child to reflect on the chain of actions he performed, on their
adequacy for the desired result, on their repeatability in other problems,
on their adaptability to slightly different circumstances, etc., the same
does not seem to be true of language activity if we take aim and result to
be linked solely to another person's reactions. This is hardly surprising,
since language is not only a means of communication, but also a system of
representation.

Another aspect of verbal activity which differentiates it from other
intelligent activities is the lack of a clear boundary between subject and
object. A child (even a baby) can hit a ball, a person or a chair, and
immediately observe the effects of similar actions on different objects and
notice the match or mismatch of the result of his activity with what he
expected from it. Objects have properties of their own, and therefore the
outcome of an action is not directly foreseeable from the action itself: a
ball thrown against a wall may indeed come back to the thrower, but it may
also bound away in a different direction due to an irregularity of the wall,
or only bound back a little way and then roll if it is made of wool, for
example. Moreover, in such actions the child may be observed to behave in
a systematic way, throwing first a ball, then a doll, then a block, etc.,
and often, from the child's postures and facial expressions of surprise,
disappointment, etc., the intention of his activity can be inferred. By
contrast, when an utterance is produced, it necessarily matches completely
the physical activity of the speech-production organs and does not have any
other properties than those supplied by the speaker himself.

Should we therefore conclude that Piaget's theory of awareness as progressing from the two first conscious aspects of every intentional act, aim and result, to conscious conceptualization starting from the most observable and exterior features of the interaction (i.e., of the subject's actions and the object's reactions, in a movement of deeper penetration in two opposite directions) cannot be relevant to the conceptualization of language? Or should we try to re-define some of the theoretical constructs of the theory of awareness so as to make them applicable to the study of conceptualization of language? Taking the latter option, the following considerations come to mind.

Once a speaker has heard himself speak, he may decide this his utterance does not correspond to the meaning he wanted to express, or that it does not correspond to certain patterns (morphosyntactic as well as phonetic) whose regularities he has noticed before, in his own and other speakers' utterances, or even that it is not in accord with what he has consciously learned at school. In this way, though in language behavior there is no observation of a discrepancy or coincidence between aim and result, some kind of matching between what one has said (the result) and a mental representation of what one should have said (the aim) does take place; aim and result in this sense are both accessible to consciousness.

This matching seems to concern both meaning and form, the two inextricably linked aspects of all utterances. The meaning has been constructed by the speaker at least partially before he starts his utterance and is essentially his own; the form the utterance takes corresponds to a mental representation he is able to constitute with the "rules" as he knows them, but these rules have been constructed through inferences from other speaker's utterances, and in many cases the final product will also be matched against a stored representation of comparable, previously perceived utterances. In other words, some parallel to the double orientation of awareness as described by Piaget can also be found in language activity, an interiorization movement concerned with meaning, and an exteriorization movement concerned with form. Following this line of speculation, a parallel with the gradual development of awareness starting from the most directly observable characteristics of the interaction between subject and object as described by Piaget can be drawn for the gradual process of conscious conceptualization in language. We may find that the first clear instances of awareness (as opposed to representation without awareness—but how are we going to draw the line?) are either reflections on what things are called or on the way certain words

should be pronounced. Further penetration into the meaning aspect would
lead to a reflection on relational meanings and further penetration into the
exterior form would lead to the discovery of phonological and morphosyntactic
regularities. Finally, a grammar may be constructed which formally links
meaning to outer form, and such a grammar goes beyond what can be inferred
from the outer form in the same way as logico-mathematical systems go beyond
the induction of regularities in physical events: it permits the theoretical
construction of sentences that can never be spoken, and at the same time it
makes for deeper understanding of the underlying structure of actual utterances. Certainly we shall find that the ultimate biological sources of
language remain inaccessible to conscious conceptualization, as Piaget argued
for all cognitive constructs.

The as yet scanty data on awareness and conceptualization in language
do not permit any clear confirmation or rejection of this rough sketch of
the possible developmental process. Some kind of matching of an utterance
to a mental representation seems to start at a very early age (see Clark in
this volume). However, as Kaper already argued (1959, p. 129 sqq), not all
"repairs" can be interpreted as self-corrections: young children often express
the same meaning several times in immediate succession with slight variations
of form, without any indication of their having noticed a mismatch; just like
adults, children often repeat themselves, for various reasons. Some "repairs"
almost certainly involve a matching of mental representations, but many are
difficult to interpret: no overt acts or facial expressions can help us here,
since young children do not look surprised or disappointed when they have
said something they are going to correct or elaborate. Apart from the cases
where the addressee was supposed to react in a specific manner, verbal acts
cannot succeed or fail in the same sense as actions on objects.

Nonetheless, there appear to be a sufficient number of cases where very
young children do indeed correct themselves, groping for a more suitable form
of expression; and on those occasions a matching against some kind of mental
representation must take place. It does not seem too far fetched to suppose
that there are indeed two kinds of representations against which an actual
utterance is tested: one of the intended meaning, and one of the morphopho-
netic form. The two must be closely linked, but sometimes testing could go
on primarily against the meaning representation, whereas at other times, the
phonetic representation may be the main point of reference. These first
matchings may well remain unconscious. When we say "first" we do not mean

to imply that these early matchings disappear at higher levels; like Piaget, we mean that these could be the first conceptualizations as well as the first "awareness" that are accessible to subjects--both for very young children in their early utterances and for older children (and adults) when dealing with specific, more difficult problems.

At later ages, but still quite early, the available data on awareness (whether observational or experimental) show that children are clearly interested in and reflect on language itself: in spontaneous behavior, remarks on word meanings, on "what one should say," corrections of other children, plays on words, etc. can easily be observed. Experimental studies become possible in which children are asked to give judgments on sentences, to break sentences up into smaller units, to explain how they succeeded in constructing certain sentences (e.g., passives, see Sinclair, Sinclair, & Marcellus, 1971) and even to reflect on some of the basic features of language (see Berthoud-Papandropoulou in this volume). Such studies have, of course, their own intrinsic interest; but if the parallel with Piaget's work on awareness in other cognitive domains holds, and I personally believe that it does, their importance is far more general: they offer a particularly fertile method of studying the mechanisms of language acquisition.

*ACKNOWLEDGEMENT*

The research on which this paper is based was carried out with the help of the Fonds National de la Recherche Scientifique Suisse, Grant No 1.190 - 0.75 and 1.527 - 0.77.

*REFERENCES*

Inhelder, B., Sinclair, H., & Bovet, M. *Apprentissage et structures de la connaissance.* Paris: Presses Universitaires de France, 1974.

Inhelder, B., & Chipman, H.H. (eds.), *Piaget and his school.* New York: Springer Verlag, 1976.

Kaper, W. *Kindersprachforschung mit Hilfe des Kindes: einige Erscheinungen der kindlichen Spracherwerbung erläutert im Lichte des vom Kinde gezeigten Interesses für Sprachliches,* Groningen: Wolters, 1959.

Moreno, L., Rayna, S., Sinclair, H., Stambak, M., & Verba, M. Les bébés et la logique, *Cahiers du C.R.E.S.A.S.,* No. 14. Paris, 1976.

Piaget, J. *La prise de conscience.* Paris: Presses Universitaires de France, 1974(a).

Piaget, J. *Réussir et comprendre.* Paris: Presses Universitaires de France, 1974(b).

Sinclair, A., Sinclair, H., & Marcellus (de), O. Young children's comprehension of passive sentences, *Archives de psychologie,* vol. 41, no. 161, 1971.

# Grammar as an Underground Process

P. Seuren

Philosophical Institute, University of Nijmegen
Nijmegen, The Netherlands

*GRAMMAR AND PSYCHOLOGICAL REALITY*

There has been a nagging problem of incompatibility in psycholinguistics between linguistic and psychological theories and results. Linguists should
be bothered by this as much as psycholinguists. That they are not I take to
be due to differences of development, scope and emphasis within the two disciplines. Linguists tend to concentrate on linguistic problems of description within a given framework (or "paradigm") and pay little need to outside
evidence or other considerations. Psycholinguists, on the contrary, tend to
take information about grammar as one of their starting points, and see how
this fits in with what they can find out about cognitive functions.

Quite a few authors have noted the conflict between transformational
linguists' dictations and the results of psychological research. I shall
mention a few, but many more could also be cited. Levelt (1974, III, p. 141)
concludes that "on the one hand the role of linguistic grammars in models of
the language user is diminishing in a way, although it is decidedly not to
disappear, and on the other hand, the theory of formal languages and grammars appears to be of increasing importance for the nonlinguistic aspects of
such models." Clearly, there is a problem here as long as we adhere to a
realistic conception of scientific theories and to the minimal requirement
of consistency of science. At the official inauguration of the Max-Planck
Projektgruppe für Psycholinguistik (May 4, 1977), Levelt remarked in his address that the capacity of human short-term memory is far too limited to
store the complex procedure involved in transformational processing from

semantic input to phonetic output.

A similar point was made by Ingram (1971). She observes (p. 337) that the transformational machinery, if taken as a model of performance, would far exceed any reasonable time limit given that "no single cognitive process takes less than one tenth of a second." Her view is "that the phenomena to which linguistic theory is applicable are essentially different from the phenomena to which psychological theory is applicable, and that therefore linguistic models and psychological models relating to language must be different; and further that the differences can be found by considering the proper place of the rules of algorithms and of heuristic strategies in each model" (p. 335). Rejecting the principle of the consistency of science, only the internal consistency of each particular science remains. The place assigned to grammar is only specified negatively: algorithms (i.e., grammars) "cannot function as models for language behavior, because they cannot function in real time" (p. 337; cf. p. 345). No answer is provided, however, to the question of where and how grammar functions if it is banned from real time.

Fodor, Bever and Garrett (1974) (henceforth FBG) put forward the interesting view (p. 241) that "experiments which undertake to demonstrate the psychological reality of the structural descriptions [of sentences] characteristically have better luck than those which undertake to demonstrate the psychological reality of the operations involved in grammatical derivations." They go on to say: "Why this should be so is a question of considerable theoretical interest." Which, of course, must be true if one considers that, barring magic, the language user must somehow convert his psychologically real deep structures into psychologically real surface structures. If transformational grammar (TG) does not provide the correct description of that process, then there must be another description of it, requiring fewer operations and complexities. But then the latter would, by all standards, be empirically superior to TG. No such alternative has been developed, however. We are stuck with TG, even though it looks as if TG is psychologically indigestible.

Marslen-Wilson (1976) concludes (pp. 225-26) that "the mental representation of linguistic knowledge cannot be adequately characterized by a transformational linguistic theory. In other words, it seems appropriate for psycholinguists to look for ways of characterizing the structure of human linguistic knowledge that are more in tune with active perceptual processes, and less constrained by the requirements of descriptivist formal linguistics."

His main argument is the fact that in current transformational theory sentences are generated from deep to surface structure, whereas many experimental results suggest that, at least in the comprehension process vigorous forces are at work which operate on a left-to-right basis. A defender of TG could answer straightaway that the left-to-right axis is known to be essential for certain phenomena in TG (pronominal anaphora, quantifier scope) (see for example FBG, 1974, pp. 214-17, and Seuren, 1969, pp. 104-19) but however that this particular dimension has not been systematically explored in TG theory. In principle, no incompatibility between psychology and TG has been demonstrated, on this score at least. TG allows for extensions and developments (possibly in the area of surface structure constraints) that could well prove to resolve the left-to-right dilemma. But, needless to say, the linguistic world has not so far shown much interest in this particular objection, just as it hasn't to most objections of this nature. And no empirically justified solution to this, or similar problems, has so far been offered.

In some cases, such as Marslen-Wilson's position quoted above, it looks as though an expansion and elaboration of linguistic theory might provide a solution. But in other cases, such as those quoted earlier (Levelt, Ingram, FBG) the opposite strategy can be put to some advantage. There a solution might be found in doing something about psychological theory. This is what I shall attempt to do in this paper. In order to help solve the problem of the incompatibility between psychology and linguistics I shall try to draw a distinction between two kinds of psychological mechanisms. The view will be put forward that there are routine procedures which escape any form of introspection and one "central control," which is (largely) open to introspection and whose operations can be brought to awareness.

## THE PROBLEM REPRESENTED IN LINGUISTICS

It should be noted that the problem at hand is as unresolved in linguistics as it is in psychology. In linguistics it is known as (or is very similar to) the distinction between competence and performance, a distinction which has proved to be both problematic and useful.

Since Katz's famous article on mentalism (1964), linguistics has been avowedly mentalistic (i.e., realistic), as opposed to instrumentalistic. The object of grammatical description, *competence*, is meant to be psychologically real. Competence must be acquired by the young child (or other language learner); a special psychological mechanism, LAD (the Language Acquisition Device) was postulated by Chomsky to account for the (still largely mysterious) processes of language acquisition. In a formal linguistic grammar items from the lexicon are selected and inserted into linguistic structures according to special rules. Linguists differ considerably as to the precise details. No one, however, in either psychology or linguistics, denies the psychological reality of the lexicon. The question of the psychological reality of the rules involving lexical items is therefore moot.

However, in linguistics no clear account has so far been given of the relation between competence and performance. Chomsky (1965, pp. 139-40) claims that it is absurd to regard "the system of generative rules as a point-by-point model for the actual construction of a sentence by a speaker." This would be "totally misconceiving its nature." No argument is given to support this view other than that "it seems absurd to suppose that the speaker first forms a generalized Phrase-marker by base rules and then tests it for well-formedness by applying transformational rules to see if it gives, finally, a well-formed sentence." *In Language and mind* (1972b), Chomsky again (pp. 116-17) asserts the absurdity of such a view, though a little more circumspectly. He proposes that the grammar should be seen as a part of the theory of performance, not as identical to it. No further specification as to how precisely a theory of competence is to be incorporated into a model of performance is given.[1] Statements concerning this question are usually negative ("we are not entitled to take this [i.e., the grammar] as a description of the successive acts of a performance model"), in spite of the explicit statement (p. 115) "that a person with command of a language has in some way internalized the system of rules that determine both the phonetic shape of the sentence and its intrinsic semantic content--that he has developed what we will refer to as a specific *linguistic awareness*."

One wonders why deep structures, surface structures, semantic representations, and the lexicon are awarded psychological reality and are declared open to psychological experiments, whereas the rules are refused the status of psychological operations. Trivializing the problem by speaking of "simple distinctions" which must not be overlooked if "great confusion" is to be avoided (Chomsky, 1972b, p. 117) is of no avail. Concealing the problem by adopting window-dressing terminology ("generative rules...may be interpreted as (purely static) conditions on the well-formedness of derivations"--Huddleston, 1977, p. 249) is no help either. The problem remains real and of considerable theoretical interest.

It is fair to say that the problem has been repressed in linguistics. This repression may well have serious consequences. Neglecting the problem is already leading to an increased isolation of linguistics as a discipline. Its closest neighbors, the psychologists, can dismiss linguistic theory as irrelevant with increasing ease. Linguistic theory, for many years a central discipline in the human sciences, may lose much of its interdisciplinary relevance.

Internally, the question is equally important for linguistic practice and theory. Although in many cases the linguist's daily work will not be affected by the problem of the psychological reality of transformational rules, the days are approaching when conflicting general theories of language or grammar will be subjected to comparative tests of psychological, or even neurological, plausibility. On the day of judgment, who will attend to the linguist's cry that his grammar is meant to be nothing more than "a system or rules that expresses the correspondence between sound and meaning" (Chomsky, 1972a, p. 62), or "a characterization of the intrinsic tacit knowledge or competence that underlies actual performance" (Chomsky, 1965, p. 140)? It simply is a fact of life that confirmation by neighboring disciplines strongly reinforces a theory, whereas incompatibility creates embarrassment.

Expressions such as those used by Chomsky to gloss over the incompatibility problem are fundamentally unclear within a mentalistic (or realistic) conception of science. The notion of *formal characterization* is, or can be made to be, perfectly clear as a description of certain properties of a hypostatized object in terms of an axiomatized system. It is then a mathematical notion. But transformational grammar is not a mathematical notion, though a formal characterization of it can be given in mathematical terms (see Peters & Ritchie, 1973). It is, on the contrary, an empirical notion. As such it is in search of an ontology.

The most natural ontology for a transformational grammar is provided by the processes of actual production and understanding of utterances. If a particular theory of grammar is psychologically absurd, or contains elements which are psychologically absurd (as the *Aspects*-model, with its base-generated phrase-markers and its awkward rules of lexical selection as presented on pp. 79-106), then this is a strong indication that that theory of grammar is seriously inadequate.

## AVAILABLE EVIDENCE AND AWARENESS

In most recent psycholinguistic literature (cf. FBG pp. 241, 273-74, 368) it is stressed that there is no clear indication of a transformational procedure being used by the speaker-hearer. However, one interesting possibility has not generally been considered. In no way does the available evidence rule out the possibility that the language user's brain runs through a transformational procedure as an automatic algorithmic program which is completely screened from introspective access or control, i.e., from awareness.

It is decidedly odd that, in linguistics as well as in psychology, the concept *psychological reality* has almost invariably implied the criterion of accessibility to awareness, in spite of the declaration, often repeated, that competence consists of *tacit* knowledge. The term *knowledge*, albeit *tacit*, may have contributed to this. Intuitively, one would like to be able to say that the point of having knowledge is precisely that it (that is, its very propositional contents) can be put to use in the complex functions of thinking and acting, both of which are to some extent open to introspection. We tend to speak of *skill*, rather than *knowledge*, in the case of functions whose underlying principles are neither accessible nor useful to awareness. Skills in this sense are called upon in relation to some purpose to be achieved. We know of the purpose, and we know we have, or do not have, the skill to achieve the purpose. But only in rare cases, if ever, can we provide an exact analysis of the skill.

The use of a language is a case in point. We know whether or not we have command of or competence in a language. When we do, we can use it for certain purposes we have knowledge of: we know what we want to say, or what the utterance is we want to understand and interpret. But the principles, or rules, by which we guide the processes of production and comprehension escape our introspection. They can only be approximated by way of scientific hypotheses and theories. All this has been said so often that one hesitates to repeat it once more. Yet it looks as though there has been a certain reluctance to accept the consequences. In the psychological literature on the reality of transformational grammars this distinction is not made in any systematic way. Ingram (quoted above) discredits TG on account of the time factor involved: no cognitive event, it is claimed, takes less than a tenth of a second. But the examples given--recognizing speech sounds or printed characters, scanning words or sentences for meaning--involve operations which require full awareness, followed by a report. It is difficult to see how this could have a bearing on processes taking place "out of sight," with no possibility of reporting on them. Such processes require time, as well as the accessible operations, but the time intervals involved are of a completely different order. It is a priori to be expected that the cognitive events mentioned by Ingram should take a lot longer than mere routine processes.[2]

An interesting case is provided by the series of experiments carried out by Miller, McKean and Slobin (MMS), as reported and discussed by FBG (pp. 227-34). MMS measured the time intervals needed by subjects to either identify or produce a target sentence which differed from a stimulus sentence in that one of the two was negative (N) and the other was not, or one was passive (P) and the other not, or one was a *yes/no*-question (Q) but the other not, or any combination of these. Sentences without either N, P or Q were called "kernel" (K). They found:

(a) a linearly additive function: "for example, sentences which involve both the negative and the passive transformations appear to require a time approximately equal to the sum of the average time required for negative or passive applied separately" (p. 229);

(b) uniform time differences according to the parameter involved: e.g., negative takes less time than passive; pairs of sentences one of which is K take less time than corresponding pairs without a K;

(c) no time difference corresponding with the direction (from stimulus to target sentences or vice verse): e.g., from K to N takes as much time as from N to K.

FBG are impressed with these results. They feel that on the basis of these results many would be inclined to accept the psychological reality of transformational rules. Yet, they then proceed to deny the relevance of these results for this question, arguing as follows (p. 231):

"We must bear in mind that transformations are ordered with respect to one another in grammatical derivations. Thus, if Miller and McKean's subjects were to apply the grammatical transformation $T$ to a given stimulus sentence in the experimental task, they would first have to recover a representation of the sentence structure as it appears in the derivation *prior* to the application of any rule ordered later than $T$. When we take this requirement into account, however, we predict relations among the sentence types that are quite different from those obtained experimentally. The clear inference appears to be that Miller and McKean's subjects could not have been using the grammatical transformations to perform the experimental task."

FBG proceed to illustrate this point by calculating the predictions which follow from this procedure on the basis of certain grammatical assumptions

about the relative ordering of transformational rules. These predictions
then turn out to be widely at variance with MMS's results. Some linguists
would take issue with FBG on the grammatical assumptions underlying their
calculations (an all too familiar pattern when psycholinguists test linguistic
theories), but that is of no relevance here. Presumably, on any ordering of
the rules the results would conflict with (a-c) above. Much more relevant
is FBG's tacit assumption that speakers should have control over decisions
to go to one particular intermediate stage in the transformational derivation,
stop there and carry on adding or subtracting one or more transformations.
As FBG correctly observe, there is no evidence at all that they do so. Nor
is there any reason to suppose that they would. If a TG forms an algorithmic
automatism, one would expect that (for the speaker) input structures are
under the speaker's control, and that the auditory output is or can be checked
(monitored), but that the processing takes place beyond his control. There
is thus no argument against the psychological reality of TG, only, it seems,
the absence of experimental evidence for it.

In fact, this picture of controlled semantic input, monitored phonetic
output (and, analogously, perceived phonetic input and recorded semantic output), the intervening processing being kept entirely "underground," fits MMS's
results rather well, provided we assume the following procedure. Let us
suppose their subjects analyzed the input sentence till the level of semantic
representation (SR), then carried out changes in the semantic contents, and
subsequently processed the new SR's into a new surface structure. The following diagrams illustrate this for K to N, and N to P, respectively.[3]

Diagram 1   $SR \longrightarrow N \longrightarrow SR'$        Diagram 2   $SR \longrightarrow \left\{ \begin{array}{c} -N \\ +P \end{array} \right\} \longrightarrow SR'$

$$\begin{array}{cc} \uparrow & \downarrow \\ K & N \end{array} \qquad\qquad \begin{array}{cc} \uparrow & \downarrow \\ N & P \end{array}$$

In this interpretation the linearly additive function mentioned under (a)
above is the result, not of any traffic up and down the vertical (transformational) arrows, but of modifications in SR's. This is compatible with (c),
provided there is no time difference between either adding or subtracting a
semantic element (N, P, or Q) in SR's.

Is there no evidence, then, confirming the status of transformational
rules as "underground" processes? It would be disquieting if this were so.
The data referred to under (b) above might well be seen as resulting from

transformational processes.  As long as the stimulus sentences are kept at a simple enough level, one would expect that both SR-modification and the grammatical processing are simpler for negation than for the passive.  This is what the data show.  It is equally predictable (though as far as I know untested) that considerably longer times will be needed for less straight-forward negatives, as in the following pairs:

(1) a.  Steve still lives in Paris.

b.  Steve no longer lives in Paris.

(2) a.  I had already left.

b.  I had not yet left.

(3) a.  Nigel is as big as Frank.

b.  Nigel is not so big as Frank.

(4) a.  Every morning he reads two poems.

b.  He doesn't read two poems every morning.

(5) a.  Most strikers have started work again.

b.  It isn't true that most strikers have started work again.

Since the SR-operation of adding a negative should not be significantly different from the "normal" cases,[4] the expected differences in processing time would be attributable to the grammatical processing.

We have, furthermore, the evidence obtained by Weigl and Bierwisch (1970). They found a particular patient suffering from aphasia, alexia and agraphia who could not write down sentences, either spontaneously or on dictation. But she could copy written sentences.  She could also, having copied a sentence, then write it down spontaneously or on dictation: the sentence was then "deblocked." Weigl and Bierwisch found that, having thus deblocked a given sentence S, the patient could, in addition, write down dictated sentences which were considered transformational variations of S, i.e., S with a preposed object, the passive of S, WH-questions where the WH-element replaced the subject or the object of S, etc.  In these cases, the patient showed some hesitation.  They write (p. 15): "Even these hesitations and trials, however, show in a very impressive way how the process is governed by the transformational network relating these sentences to the common underlying structure." Although it is not immediately clear how the details of these findings should be interpreted, it does look as though they are in good agreement with the notion of grammar as an underground process.

But apart from all clinical or experimental data, there is the whole of linguistic theory, insofar as it has been argued for on "linguistic" grounds, providing evidence for the reality of transformational processes. The chief difference between a linguistic and a psychological argument consists in the fact that, in linguistics, data collecting is less systematic and notionally less clear, whereas theory seems more developed than in psychology. But if linguists were more conscientious in registering native speakers' linguistic attitudes (following Labov's, 1975, admonitions), and if the categories employed (*grammatically well-formed, acceptable, semantically well-formed, substandard, dialectal,* etc.) were given a more solid basis, few psychologists would maintain that "experimental investigations of the psychological reality of grammatical rules, derivations, and operations...have generally proved equivocal " (FBG, p. 368).

Given that transformational processes are screened from introspective access, there is little else one can do but approximate them by hypothesis, i.e., theory. Suppose, for example, that it has been established experimentally, under controlled conditions, that speakers of English (of either all or only some social classes or geographic areas) accept and use sentences such as (6) and (7)

(6) a. Anxious as Edith always was to please her son, he grew up a spoiled child.

    b. Reluctant as I was to leave, I kept holding up my glass.

(7) a. Difficult to erase as it was, the slogan remained visible for a long time.

    b. Easy to fool as Henry was, he soon lost all his money.

    c. Hard to persuade as Jimmy is, he'll have to suffer.

much more naturally than sentences with the systematic difference between (6) and (7) inverted ("Anxious to please her son as Edith always was,....; difficult as it was to erase, the slogan...."). Then it is difficult to imagine anyone objecting on scientific grounds to the inference that there is, in some way, a stronger cohesion in constructions of the type "easy to fool" than of the type "anxious to leave." Given a sufficient amount of specialized theory, a linguist might then propose different constituent structures for both classes.

Suppose further that it has been established experimentally that subjects provide close-enough synonyms for particular combinations involving adjectives of the (7)-group[5] with much greater ease and frequency than for those of the (6)-group. In the light of bits of existing theory of lexical formation, one might then feel justified in postulating that (7)-type collo-

cations are dominated by a single categorial node (in this case "Adjective"),
as in Diagram 3, whereas (6)-type constructions have a structure as in Diagram 4.

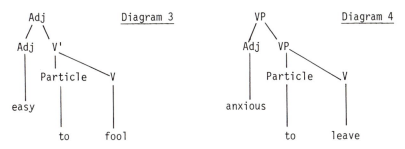

The theory might involve the assignment of special, definable, properties to
subtrees dominated by a categorial node (as in Diagram 3), such as the immediate substitutability by a single lexical item, or the relatively strong
resistance to interruption (=cohesion).

Whether theoretical statements such as these are, or are not, correct,
useful or fruitful, is a question to be answered primarily by the specialist
in linguistic structures, namely the linguist or grammarian. He, however,
has no reason to be distrustful, or even wary, of experimentally obtained
data (as has become the prevalent attitude in linguistics[6]). Likewise, the
psychologist has no reason, as far as I can see, not to accept the theory of
grammar as an integral part of psychological theory. This applies to the
linguistic structures as much as to the rules. One cannot accept evidence
about the reality of deep and surface structures without postulating transformational processes occurring between them. If they are beyond introspective reach, as we have to admit they are, the simplest theory, i.e., the one
capturing the widest generalizations with the least apparatus, will count
as the strongest.

*TWO MECHANISMS*

The claim that grammar is an underground process makes little sense if
one is not prepared to postulate two kinds of cognitive machinery, one being
of a more routine-like nature, the other having an integrated central control
function. For lack of better terms I shall speak of *routine procedure* (RP)
and *central control* (CC), respectively. Since the distinction is not intuitively repellent and also seems to have some explanatory potential, I shall
try to formulate what seem to me to be necessary minimal conditions for the

two kinds of mechanism.

Let us suppose that RP's are characterized by the fact that they consist of self-contained algorithmic programs of a non-recursive nature, not sensitive to independent variables other than the inputs received. The input is processed into an output without any outside interference. Once the program has grown or has been acquired, and given the right physical condition of the organism, the only condition for a proper functioning is the feeding in of the right kind of input.

The other kind of machinery, CC, receives inputs from various RP's and provides outputs for and instructions to other RP's. It operates throughout in terms of stylized representations. These cannot be a "language" in the linguistic sense of the word, i.e., tree structures subject to certain strictly defined grammatical constraints. Fodor (1975, pp. 55 ff.) is very clear on this (although I regret his use of the term *language* for thought-structures: in this context the term is an unhappy metaphor). They will no doubt be as severely constrained as linguistic structures, but in entirely different ways. The important thing here is that they are non-linguistic. We shall speak of them as *cognitive representations* (CR), as opposed to semantic representations (SR) which are linguistic structures. No attempt will be made here to specify more precisely the structural or functional principles of CR's. It must be mentioned, however, that the notion is as old as, and largely identical with, that of proposition (see Nuchelmans, 1973). Inasmuch as the traditional notion of proposition is meant to refer to non-linguistic cognitive representations (thought-structures) underlying sentences, it coincides with what we call CR here.

Routine procedures have been described as having an input and an output. Since their main function and *raison d'être* is to do the groundwork for central control, it is reasonable to suppose that CR's, which are the typical representations of CC, are always involved one way or another. So we stipulate (provisionally) that RP's operate from given inputs (possibly sensory data) to cognitive representations, or from these to outputs of a different nature (such as linguistic outputs), or from CR's to CR's.

We want, of course, central control to be the part of our mental machinery that we "have access to." We are, or can make ourselves, aware of what goes on there, and we can give verbal reports (*protocols*) of that. In fact, that is what is meant by *having access*. However, this impressionistic parlance sounds as though somebody inside us, a "homunculus," is keeping an eye on what happens in CC. This, of course, cannot be literally so, since that little man would again have his mind, with a CC and an even smaller homunculus in it, and we would be led into an infinite regress. On the other hand, there does seem to be some recursion. I can make myself aware of the fact that I am aware that I am slowly getting drunk. Ideally, though not in practice, this could be carried out *ad infinitum*. Only limitations of storage and processing capacity seem to keep us from going up too high in the awareness spiral. In other words, there must be some recursive mechanism in CC.

An obvious way to account for this form of recursion is to remember that we have a memory. Our memory, needless to say, is very unlike the kind that computers are equipped with. What exactly it is like, is still very much an unresolved question, but enough seems to be known to venture the inference that it must be structured according to categories, propositions, theory-like sets of propositions, unifying generalizations. The basic element of human memory seems to be the proposition, or as we have dubbed it, CR.[7] There also seems to be a gradual downward movement in our memory system, from isolated *ad hoc* occurrences which are kept for very short periods only, to integrated propositional complexes which have the structure of deductive theories and are kept for much longer stretches of time (according to some, forever).[8] Only some data seem to pass into longer term strata of the memory. So one way or another, there must be a filtering procedure. As the memory gets deeper, less is known about its organization. In particular, we must confess to almost total ignorance of the ways and means by which items sink deeper and deeper into memory while they undergo gradual transformation so that they are, in the end, encapsulated into integrated cognitive complexes. The deeper strata are usually called *knowledge*.[9]

But apart from the other mysteries of memory, there is one assumption that has to be made a little more explicit here. Let all events occurring in CC be recorded in memory, and only these. Let CC have the possibility of putting a special kind of RP into action, which will search memory for certain "information" and bring it back to CC. Then, the instruction given to the search procedure is itself kept in the memory, and CC can therefore switch on the search procedure again to get the previous instruction, and so on *ad infinitum*.

If the information to be retrieved from memory is at its freshest, i.e., if a CC-event is being, or has just been, recorded without interruption, the search procedure will stay at its minimum: it will immediately "report" (and the report can be verbalized) that recording is going on. But this report is again recorded, and that process can be reported on again, and so on. I believe that this immediate reporting of the recording of current CC-events is a fairly close approximation of what we call *bringing to awareness*.

If it is true that all CC-events, and only these, are recorded in memory, they must be extremely costly in terms of brain capacity (and probably also in terms of time). It certainly makes a great deal of sense to postulate routine mechanisms that are less costly and do the jobs which have to be done but whose step-by-step procedure is irrelevant for the decisions to be taken by the organism as a whole. There is no point in storing these procedures in memory. Quite clearly, it is a matter of good functional economy that RP's, which are nothing but ancillary routines, are screened from access so that they cannot be interfered with by some CC-command, retrieved or reported upon.

It is quite thinkable (and in the case of grammar probably true) that a given complex RP has the following property. For each input there are a number of successive stages, $s_0$, $s_1$,..., $s_n$, it has to go through, where $s_0$ is the input, and $s_n$ the output of RP. However, there are $s_i$'s ($0 < i < n$) such that the transition (by rule) to $s_i + 1$ is contingent not only, or not at all, upon the structural properties of a $s_i$, but (also) on one or more previous stages. RP will then have some means for keeping "tabs" for the duration of the processing of $s_0$ into $s_n$. A psychologist who likes computers would now say that this RP is a system with a memory attached to it, since it has this tab-keeping device. Yet it should be clear that the term *memory* is not applicable here. CC has no access to these tabs. No reporting is possible on them. And there is no reason at all to suppose that they have a propositional structure. When I speak of memory, I wish to refer specifically to that faculty in our minds which enables us to remember things, and not to just any cognitive functions causing storage of any kind to take place.

It often occurs that a particular activity, which was learned explicitly and in full consciousness (the learning taking place, so to speak, under direct supervision of CC), becomes an automatism after a while and with practice. This phenomenon, which is extremely well-known both in the literature and in each individual's personal experience, implies that it must be possible for particular classes of occurrences in CC (most probably not for all kinds of CC-events) to be degraded to the status of RP. Something of this nature seems to happen when adults learn a foreign language by an explic-

it method. If, and to what extent, some such process also occurs during
first language learning in early childhood, is an interesting question.[10]
Other examples are car-driving, skating, bicycle-riding, but not, apparently,
the making of sums.

Unlike RP's, we must assume that CC is under heavy pressure from all
kinds of independent variables. This is more or less implied by our saying
that CC receives inputs from various RP's feeding into CC, and that some of
these RP's receive their inputs from outside (sensory data, sensations, emo-
tions). To the extent that CC is influenced by the inputs it receives, it
is subject to independent variables. I think that it is the interference,
occurring with greater or lesser intensity, brought about by various RP's
feeding into CC, which determines whether or not CC will switch from one
chain of operations to another (shift attention, that is). This, of course,
takes us right in the middle of a very old and extremely complex problem:
what, if anything, is *free will*? I do not claim to know the answer. But
I do claim that it is possible to construct a model where the decision regarding the kind of activity that CC will be occupied with next is a function
of a complex interference pattern of RP's feeding into CC. What we are going
to think about next, in other words, and also, therefore, what we are going
to do next, is determined by the intensity with which certain RP's deliver
their product to CC, in relation with the strength of other programs being
planned or carried out.

Constructing an adequate model of CC will require a great deal of theory,
so far non-existent. Among other things, such a model will have to specify
degrees of intensity and strength, as indicated above. It will have to
specify the form of CR's, the kind of operation that can be performed on
CR's, the way or ways in which a definable set of operations can become an
automatism, and many other things. But in a way, and from a philosophically
safe distance, these are technical problems. There still remain the philosophical problems raised in certain quarters, especially the philosophy of
personal identity, that of mind, and the philosophy of action. Does our
distinction between CC and RP's provide the possibility in principle to cap-
ture the elusive *I* so much discussed in the philosophical literature?[11] For
example, do *I* have control over the relative intensity with which *my* RP's
deliver their output to *my* CC? If so, what does this mean for the mechanism
that I am, or have?

One question which has bothered philosophers engaged in the body-and-

mind problem (or the dualist-materialist issue) is presented by the fact
that the word *I* can be replaced by *my body* in some cases but not in others
(and, analogously, *you* can sometimes be replaced by *your body*, *he* by *his
body*, etc.). Thus, in expressions such as *I am heavy, I am still alive, I
was hit by a bullet, I am turning grey*, the word *I* can be replaced by *my
body* (or, as the case may be, by an expression denoting a part of the body,
such as *my hair*). But when I say *I am sad, I am reading, I like ships*, the
same substitution is out of place.[12] Does one not detect some underlying
primordial dualism here? How is this to be reconciled with the monist-
materialist view of man which has been current in psychology at least since
the advent of behaviorism? There is, it seems to me, some perhaps considerable mileage in the position that if we oppose body and mind, taking *body*
in a narrow sense, then the term *body* stands for all RP's plus all the other
non-cognitive routine processes that keep us going. *Mind* then stands for
what we have called CC here. *Body* in the wider sense embraces both *body* in
the narrow sense and *mind*. When substitution of *my body* for *I* is inappropriate, this is no more than one of those cases where an expression denoting a
part of the body must be used, in this case *mind*. The conclusion will then
be that there was, after all, just a confusion between the two senses of
*body*, and the dispute was, after all, based on a terminological equivocation
(see footnote 12). The real answer, in the end, is monist-materialism. I
do not know to what extent this line of reasoning will be tenable eventually.
But the answer will depend largely on what CC can do to help us understand
the ways in which we talk about ourselves.

Questions such as these may seem remote to a psychologist. Yet they
are real and will have to be answered. The distinction between CC and RP's
implies an acceptance of questions such as "Can a machine think?",[13] and it
implies an attempt at forging an answer. But it is not primarily such philosophically tainted questions that motivate the distinction. There are much
more direct, empirical reasons for it. One reason we have given: it helps
to solve the incompatibility problem of transformational grammar and psychological data. But, of course, such heavy machinery needs stronger, and more
general, support. And this it finds in the following general consideration.

Behaviorism has had the great merit of setting the stage for an exact
formulation of the problem of the causation of human behavior. It also
claimed to provide an adequate answer which involved absolutely minimal
assumptions about the organism. The behaviorist effort consisted mainly in

correlating stimuli and responses (behavior). The variables determining behavior were placed in the outside world, not, or as little as possible, in the organism. This position has, by and large, now been abandoned. Occam's razor was too sharp; we have had to enrich our assumptions about the organism. Chomsky (1959) neatly demonstrated the weakness of behaviorism with respect to linguistic behavior, but his arguments can be generalized to most or all forms of behavior. Only some stimuli correlate with a well-defined class of responses. But in many, if not most, cases we have an inescapable intuition that there is a correlation between this stimulus-token and that response-event even though we can give no clear behavioral definition of the class of responses associated with that stimulus.

To give an example, let us imagine a town where open civil war is raging. After a period of relative calm there is a rapid succession of very loud and very sudden diffuse noises originating from the Northwest area of the town and heard all over. There is a variety of responses. Some taxi-drivers suddenly turn left, others turn right, others stop. Some people, in houses relatively remote from the source of the noises, suddenly open their windows. But others, closer to the bangs, run out of their houses into makeshift bunkers. Armed men on rooftops start aiming and preparing their mortars, and so forth and so on. Clearly, there is no behaviorally definable class of responses to what we interpret as the series of explosions. Whatever predictions we can make about people's responses will depend on what theory we hold of the situation generally, of each person's position in the general situation, and of each person's ideas of life and death, safety and danger. Our predictions may well be rational, but any attempt at giving an explicit account of their rationality will necessarily have to make strong claims about the various responding organisms. The argument extends to long-range effects of a given stimulus. It is normal for us to detect the effects of a stimulus on a person's behavior over a period of time. Apparently, we are able to select those elements of behavior which correlate with the stimulus, even though there is not a shadow of a behavioral criterion. We do so on the basis of certain implicit assumptions about the other person's "organism." Any attempt at predicting and selecting such effects scientifically will have to make those assumptions explicit.

The absence of behaviorally definable classes of response-events (provided we accept intuitions about relatedness of stimuli and responses), implies the existence of some central clearinghouse where inputs are brought

together, processed and integrated into structures that determine elements
of behavior spread over large numbers of behavioral categories and often
over extended periods of time. In other words, a preliminary problem analysis regarding the causation of human behavior quickly shows that there must
be a powerful integrating mechanism relating stimuli with vast varieties of
responses. An explicit account of this mechanism will be an account of what
we intuitively call "rationality." A rejection of this argument will lead,
and has led, to an intolerable impoverishment of psychological theory.

The introduction of a CC-mechanism is precisely a recognition of the
reality of certain stimulus-response correlations which can only be understood on the assumption of some rationality in the organism involved. Since
behavior caused by CC-intervention is not limited to any precisely definable
behavioral category or class of categories, the laboratory is not the ideal
place to test organisms on this score. Under laboratory conditions, behavior
has to stay within clearly defined boundaries. Yet no one could seriously
impose the condition on scientific work that it should *per se* be restricted
to the intrinsic conditions and the material possibilities of a laboratory.
A great deal of scientific work consists in constructing empirical theories
(i.e., theories which can be evaluated on account of their explanatory power)
on the basis of data that cannot be gathered in a laboratory.

A distinction between processes that are and that are not accessible to
introspection has been proposed here as a way of resolving the incompatibility
problem in psycholinguistics. Innocent as it may have looked at the outset,
in fact it turns out to involve bold theoretical claims about the organization of the human mind. It has been the purpose of this paper to show (a)
that under the CC/RP-distinction there is, so far, no risk of incompatibility
between the theories and findings of linguistics and psychology, and (b)
that the distinction is needed anyway on general and independent grounds.

Some qualification is necessary, however, with regard to the statement
that under the CC/RP-distinction linguistic and psychological theory are
not incompatible. The statement is much more obviously true for that development in linguistics known as generative semantics or semantic syntax, and
much less so for Chomskyan autonomous syntax. In semantic syntax, the deep
structure is identical with the SR, which can be considered the result of
a mapping from a given CR (where the CR is a non-linguistic and the SR a
linguistic structure). The transformational rules turn SR into a surface
structure. This process is quick, uncontrollable by any non-grammatical

factors, and is not recorded in memory (although some auxiliary tab-keeping may be involved). The grammar is merely a processing mechanism operating on an unbounded set of SR's, whose origin is postulated in CC. This model of grammatical description is not generative, but only transformational.[14]

In autonomous syntax, on the contrary, SR's result from special semantic interpretation rules operating on surface structures. The deep structures of the syntax are "generated" by the rules of the base component. As we have seen, Chomsky maintains that it is absurd to suppose that a speaker first builds up a deep structure through base rules, and then runs the structure through the T-rules to see if a well-formed surface structure is yielded (1965, pp. 139-40). Chomsky is probably right here: such a procedure sounds highly implausible. His base rules are not "generative" in this sense, but are mathematical specifications of the well-formedness conditions of his deep structures. Similar conditions can be formulated for SR's, as for any class of tree-structures. But if deep structures are psychologically real, they must have a psychological origin. Thus the question of the psychologically real generation of the Chomskyan deep structures is still wide open. And it is hard to see how this question could be answered without the deep structures being mappings of thought structures, i.e., SR's. If, on the other hand, deep structures are not psychologically real, one wonders what sense it makes to say that a child must acquire the grammar of his language, or, more generally, that a grammar describes "a mental reality underlying actual behavior" (Chomsky, 1965, p. 4). In short, the whole fabric of Chomskyan autonomous syntax suffers from psychological implausibility. And, not surprisingly, this particular branch of grammatical studies also suffers from fundamental unclarities regarding such basic methodological issues as the correspondence between the postulated theoretical entities and the object of investigation. When I say, therefore, that under CC/RP-distinction there is, for the moment, no fear of incompatibility between psychology and linguistics, this applies to semantic syntax, rather than to autonomous syntax.

*FOOTNOTES*

1. FBG, pp. 370-71, distinguish five ways in which grammar can be thought to interact with a model of performance, and there is no indication that their list is exhaustive. These five ways, moreover, are so general in their own ways that many further subdivisions could be made.

2. See Clark and Clark (1977, pp. 56 - 81) for some experiments that are relevant to processing time of grammatical automatisms.

3. In accordance with dominant opinion in TG I assume that there are semantic differences between corresponding active and passive sentences, even though current semantic theory does not provide us yet with the apparatus to describe these differences systematically.

4. In the pairs (1) and (2) the SR-operation of negativizing might be argued to be more complex than in the "normal" cases, since not only does a negation have to added, but a presupposition has to be taken into account, as in (1), or modified, as in (2).

5. In fact, there are numerous examples of this type of correspondence in English, as in many other languages: *amiable* (easy to like), *handsome* (good to look at), *touchy* (easy to offend), *fragile* (easy to break), *readable* (nice to read), *gullible/credulous* (easy to fool), *light* (easy to digest), *stubborn/obstinate* (hard to persuade), etc., etc.

6. McCawley is a clear exception to this rule. In a recent interview (Aarts, 1977) he declares his interest in experiments conducted by Fodor at MIT, of which he says (p. 244) that they "may very well be very productive with regard to justifying particular analyses or to showing that certain supposed similarities aren't quite as much of similarities as they have been taken to be." McCawley also holds the opinion, still rare among linguists, that no strict separation can be made between competence and performance: "that to make sense out of what a person's knowledge of language consists in, you will have to talk about language processing: what goes on in the production and comprehension of language" (p. 240).

7. See Anderson and Bower (1974, pp. 86-87), Clark and Clark (1977, p. 136) on the propositional nature of (long-term) memory.

8. The distinction usually made between short-term memory and long-term memory (e.g., Clark & Clark 1977, pp. 135-41) does not seem to capture the whole truth. Instead of this dichotomy, it looks much more as if there is a structure causing either gradient or step-by-step transitions from short-

term to long-term memory representations.

9. In fact, memory is some kind of induction mechanism. It certainly shares many of its mysteries with induction. If the conditions for induction (i.e., for theory formation) are not fulfilled, the integration process does not work and the items disappear from memory. In such a case, the items are unrelated and fail to fall into a meaningful pattern. In order to remember such items we have to resort to memorization, and memory does not seem to like that very much. See Clark and Clark, 1977, pp. 134-35, 137, 143.

10. See, e.g., Kaper (1959), Read (1973), Slobin (this volume).

11. Cf. Wittgenstein, *Tractatus*, 5.631 - 5.641; Strawson (1959, pp. 87-116).

12. According to some philosophers (Carnap, Quine), such a question is uninteresting, since it is language dependent. Others (Ryle, Cohen) reject the language dependency of philosophical questions. See Gyekye (1977) for an interesting comparison with the Akan language, where a dualistic philosophy seems to manifest itself in expressions of the type discussed here.

13. It is remarkable that in the branch of psychology where model construction is taken most seriously, in artificial intelligence, this kind of problem is avoided or rejected. E.g., Hunt (1975, p. 444) does not accept questions of the type "Can a machine think?", and falls back on a theoretically minimal position which involves an agnosticism with respect to the materialist-dualist issue, and a rejection of the criterion of AI-programs that they simulate psychological performance. The cost of this stance is a considerable reduction of the scientific relevance of AI work.

14. For a systematic exposé of the difference between autonomous and semantic syntax, see the introduction edited by Seuren (1974).

*REFERENCES*

Aarts, F.G.A.M. An interview with Professor James D. McCawley. *Forum der Letteren*, 1977, 18, 231-51.

Anderson, J.R., & Bower, G.H. *Human associative memory*. Washington, D.C.: John Wiley & Sons, 1974.

Chomsky, N. Review of Skinner (1957). *Language*, 1959, 35, 26-58.

Chomsky, N. *Aspects of the theory of syntax*. Cambridge, Mass.: M.I.T. Press, 1965.

Chomsky, N. *Studies on semantics in generative grammar*. The Hague: Mouton, 1972(a).

Chomsky, N. *Language and mind*. Enlarged Edition. New York: Harcourt Brace Jovanovich, 1972(b).

Clark, H.H., & Clark, E.V. *Psychology and language: An introduction to psycholinguistics*. New York: Harcourt Brace Jovanovich, 1977.

Fodor, J.A. *The language of thought*. Hassocks, Sussex: Harvester Press, 1975.

Fodor, J.A., Bever, T.G., & Garrett, M.F. *The psychology of language: An introduction to psycholinguistics and generative grammar*. New York: McGraw Hill, 1974.

Gyekye, K. Akan language and the materialist thesis: A short essay on the relation between philosophy and language. *Studies in Language*, 1977, 1, 237-244.

Huddleston, R. In defence of parasitic base structures. *Studies in Language*, 1977, 1, 245-254.

Hunt, E.B. *Artificial intelligence*. New York: Academic Press, 1975.

Ingram, E. A further note on the relationship between psychological and linguistic theories. *IRAL*, 1971, 9, 335-346.

Kaper, W. *Kindersprachforschung mit Hilfe des Kindes: einige Erscheinungen der kindlichen Spracherwerbung erläutert im Lichte des vom Kinde gezeigten Interesse für Sprachliches*. Groningen: Wolters, 1959.

Katz, J.J. Mentalism in linguistics. *Language*, 1964, 40, 124-137.

Labov, W. *What is a linguistic fact?* Lisse, Holland: De Ridder, 1975.

Levelt, W.J.M. *Formal grammars in linguistics and psycholinguistics*, 3 vols. The Hague: Mouton, 1974.

Marslen-Wilson, W. Linguistic descriptions and psychological assumptions in the study of sentence perception. In: R.J. Wales & E. Walker (eds.), *New approaches to language mechanisms*. Amsterdam: North Holland, 1976.

Nuchelmans, G. *Theories of the proposition: Ancient and medieval conceptions of the bearers of truth and falsity*. Amsterdam: North Holland, 1973.

Peters, S., & Ritchie, R.W. On the generative power of transformational grammars. *Information Sciences*, 1973, 6, 49-83.

Read, C. Children's judgments of phonetic similarities in relation to English spelling. *Language Learning*, 1973, 23, 17-38.

Seuren, P.A.M. *Operators and nucleus. A contribution to the theory of grammar*. Cambridge: Cambridge University Press, 1969.

Seuren, P.A.M. (ed.), *Semantic syntax*. Oxford: Oxford University Press, 1974.

Skinner, B.F. *Verbal behavior*. New York: Appleton-Century-Crofts, 1957.

Strawson, P.F. *Individuals: An essay in descriptive metaphysics*. London: Methuen, 1959.

Weigl, E., & Bierwisch, M. Neuropsychology and linguistics: Topics of common research. *Foundations of Language*, 1970, 6, 1-18.

Wittgenstein, L. *Tractatus logico-philosophicus*. London: Routledge and Kegan Paul, 1922.

# On the Mechanics of Emma

John C. Marshall
Interfaculty Group on Speech and Language Behavior, University of Nijmegen,
Nijmegen, The Netherlands and

John Morton
M.R.C. Applied Psychology Unit, Cambridge, Great Britain, and
Max-Planck-Gesellschaft, Projektgruppe für Psycholinguistik
Berg en Dalseweg 79, Nijmegen, The Netherlands

Normal adults can and do reflect upon their own (and other people's) linguistic behavior. Although we do not wish to belabor this point, it is as well to distinguish from the outset between two types of "awareness" that must be explicated in any account of such metalinguistic abilities. At risk of seeming contradiction, we shall call the first type *tacit awareness*, and the second *explicit formulation*. In order to illustrate the distinction consider the following "sentence":

The birds is coming.

On the assumption that the advertising agency which publicizes the works of Alfred Hitchcock knows its job, we may presume that a substantial number of people found this sentence amusing and memorable. It is difficult to see why this should be so unless they were tacitly aware of the fact that either the verb should be *are* or the noun phrase should be enclosed in inverted commas. When we turn, however, to explicit formulation, we would guess that fewer people have the ability to remark that the subject of English sentences must agree in number with its verb; and, indeed, until 1955 only one person formulated a fully generative rule capturing this generalization for the infinitude of well-formed English sentences.

If people were unable to evince evidence of tacit awareness there would

be no informants; if no one was able to express the generalizations underlying regularities of usage there would be no grammarians.

*A Putative Paradox*

Such "awareness," either tacit or explicit, seems to involve having (at
least partial) access to the internal structure of the mechanisms which underlie knowledge and behavior. It is in this sense that awareness of structure (or function) is a *meta*-ability; and by virtue of being a meta-ability
the notion can easily lead to paradox if it is invoked as part of the explanation of primary linguistic abilities. Thus Chomsky (1965) draws upon the
analogy between the child's (tacit) discovery of the grammar of the language
to which he is exposed and the linguist's (explicit) formulation of that
grammar, but he is careful to stress that the child acquires this skill un-
consciously: "It seems plain that language acquisition is based on the child's
discovery of what from a formal point of view is a deep and abstract theory--
a generative grammar of his language--many of the concepts and principles of
which are only remotely related to experience by long and intricate chains
of unconscious quasi-inferential steps" (Chomsky, 1965, p. 58).

The obvious moral to draw from Chomsky's analogy is that although making linguistic intuitions explicit may be useful for the adult who wants to
write down a grammar on paper, this is clearly not essential to the child
who is writing a grammar on brain cells. It is Chomsky's biologism--"we may
regard the language capacity virtually as we would a physical organ of the
body" (1976, p. 1)--which forces him into an epiphenomenalist position. In
a particularly clear expression of this view, he continues (p. 23): "I see
nothing surprising in the conclusion, if it proves correct, that the principles of rule organization that underlie the *WH-island* constraint are special properties of the language faculty, just as distribution of orientation
specificities is a special property of the visual cortex." This position
may indeed be correct, but once such a biologism is embraced it is difficult
to see how linguistic  intuitions  could aid in the development of the language faculty any more than cardiac intuitions could aid the embryological
differentiation of the heart.

Much the same point can be made from the psychological literature. Consider the famous exchange reported in Brown and Bellugi (1964, p.135):

Interviewer:  Now Adam, listen to what I say.

Tell me which is better . . . some water or a water.

Adam: Pop go weasel.

When quoted, this exchange is usually prefaced by some such remark as "Unfortunately, we cannot obtain grammaticality judgments from a two-year old." It is important to realize, however, that the "unfortunately" refers to the state of the adult investigator, not of the child. Little Adam is not having language-learning problems; rather big Brown is having grammar-writing problems.

The point is one of logic rather than fact. If a two-year old could tell us that *some water* is grammatical but *a water* is not, that would simply demonstrate that he had already (i.e., previously) acquired the distinction between count nouns and mass nouns (or, minimally, that he had heard adults say *some water* more frequently than *a water*). First you learn something, then you *may* be able to verbalize explicitly what you have learned. But it is difficult to see how this order could be reversed. How could one have intuitions about mass versus count nouns and then learn that there is such a distinction?

Seen in this light, linguistic intuitions have the status of an optional extra, completely disconnected from the mechanisms responsible for the acquisition of linguistic skill. The "extra" has the added disadvantage of being associated with talk of awareness, consciousness, and the ego. Given no useful role to perform, linguistic awareness becomes not so much the ghost within the machine as the ghost outside the machine. We seem to be stuck with the position that William James (1890) captured so neatly with the following analogy: "So the melody floats from the harp-string, but neither checks nor quickens its vibrations; so the shadow runs alongside the pedestrian, but in no way influences his steps."

## A Change of Metaphor

In an attempt to bring linguistic awareness back *within* the system, we suggest the following analogy: Represented within many perfectly familiar machines is a distinction between *use* and *mention*. The machines we are thinking of are those in which one component can monitor the state of another component. Imagine a motor car in which a red light (on the dashboard, say) flashes when (if and only if) the petrol tank is almost empty. Such a signal *mentions* the state of the petrol tank, but is not part of the mechanism whereby the use of petrol is involved in propelling the car. To the question "Can you have (i.e., is it logically possible to have) a well-functioning

car which does not include such a monitoring device?", the answer is clearly
"Yes." To the question "Is it *sensible* to have a car without such a device?",
the answer may well be "No."

We shall rely on variants of this analogy to elucidate the role of linguistic awareness in the processes of language comprehension and production.
In our (restricted) usage of the term, then, the primary functions (and perhaps the ontogenetically earliest functions) of linguistic intuitions are
negative. Awareness arises out of devices for finding faults. In complicated
machines (e.g., devices of the complexity required for producing and understanding utterances in a natural language) many qualitatively different types
of process can go wrong. When something does go wrong, it is often helpful
if the machine can signal not merely *that* a malfunction has occurred, and
*where* it has occurred, but also *what kind* of malfunction it is. We are visu-
alizing, then, a hierarchy of monitoring, control, and repair processes which
we believe correspond with one way of using the terms *linguistic intuition*
and *linguistic awareness*.

*Building a Set of Levels*

Let us start from the premise that normal language processing proceeds
without awareness. If this were not so, then speaking would itself be evidence for (and co-extensive with) awareness, and the term would become totally vacuous. Although there is a hint of circularity in possible defenses
of this premise it seems uncontentious to agree that simple, fluent asser-
tions--for example, "That's a cow"--can be uttered without awareness of any
of the structural parameters which enter into the description of the remark.
We choose to represent normal language processing as normal language processes (NLP) in a (no more opaque than usual) black box. This is a nota-
tional idiosyncracy which we see as value-free and non-committal. For the
moment, the contents of NLP are mysterious apparati.

$$\boxed{\text{N L P}}$$

Normal language processes serve two functions. They receive inputs and
*compile* or interpret them by following a program (a program determined in
part by embryological growth, in part by environmental action). The program will be subject to modification by learning and in light of information
coming from other processes which deal with non-linguistic aspects of the
internal and external world. The NLP also cause speech to be produced, either

as a result of their own operations or by virtue of stimulation from other external or internal events. Naturally we do not assume that input and output processes are identical (indeed we believe that in some cases they are almost independent). Nonetheless, for present purposes we can begin without differentiating them.

We now note that there are some clear and uncontroversial cases of linguistic awareness, for example, explicit comments on the form of linguistic rules. Such behavior seems to presuppose the monitoring or observing of primary linguistic processes (or, better perhaps, the monitoring of the *results* of such processes). We will represent this monitoring by another box containing an even more mysterious apparatus--or EMMA. And we shall claim that EMMA can both monitor NLP and, if necessary, change the mode of operation of NLP.

$$\boxed{\text{N L P}} \Leftrightarrow \boxed{\text{E M M A}}$$

For purposes of the argument we thus define "awareness" as EMMA-functioning. Certain consequences follow. For example, in a particular situation, either EMMA operates or she does not. There are therefore no *degrees* of awareness, although the extent and type of information about NLP to which EMMA has access will, of course, vary. A further consequence of our formulation is the avoidance of an infinite regress of awareness of awareness of awareness.... Two-way communication between EMMA and NLP implies that the operation of EMMA will sometimes result in a change in NLP to which EMMA again has access. That is, what looks like awareness of awareness is simply awareness of the consequences of awareness. We hope, in this way, that the concept of EMMA may eventually remove some of the philosophical problems which arise when the term *aware* is given quasi-spiritual connotations.

*Who is EMMA?*

Here, as everywhere, we prefer to hedge rather than last ditch. We feel it will be wiser not to attempt a rigid definition of EMMA-functions; we shall rather approach her by stealth. We shall look at speech production and perception separately, attempting to specify the kinds of mechanisms that are necessary to account for phenomena which have been interpreted as evidence of awareness.

The comprehension of speech requires (at least) the conversion of an acoustic signal to a semantic code. We shall call the mechanism that accomplishes this feat the compiler, and thus represent speech recognition in the following way:

Speech input ──────→ Compiler ──────→ Semantic representation (SR)

In a similar oversimplified fashion, we shall regard language production as a system with two primary components. In the first place an expressible intention must exist which can be formulated in a semantic representation. The semantic representation must then be converted into a form which can be (physically) expressed, viz:

Formulator ──────→ Semantic representation (SR) ──────→ Expressor ──────→

Since the semantic representations (SRs) of production and recognition must match, we can link the two systems as follows:

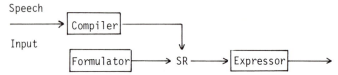

If we think of these devices as the primary components of the performance system, then any device which has access to the results of their computations without itself being part of the primary performance machinery is a part of EMMA. Let us now attempt to be a little more specific.

*Input Monitoring*

We have referred to the central component of the language understanding system as a compiler which takes speech as input and produces a semantic representation as output. Clearly, however, the infant will frequently find himself in a situation where either no semantic representation or a drastically limited representation becomes available even after a fairly long stretch of speech has been heard. The utterance "Desist from procrastination!" is unlikely to mean very much to a one-year old. If the compiler produces literally no output, then nothing further need to be said or done. If the compiler produces a very restricted output then little will be done. (One might argue that, for the infant, happiness is a warm intonation contour, and can trigger a smile.) But eventually the child does come to notice that something has gone wrong. There has been an input to the compiler and no (or too little) output. At this point we do seem to need a monitor (M), viz:

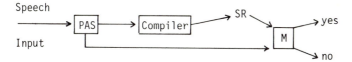

We can use an output triggered by a precategorial acoustic system (PAS), an independently motivated component of the perceptual system (Crowder & Morton, 1969), as one of the inputs to the monitor.

In their simplest form, the rules for this system are: If the inputs to the comparator are either 0,0 or 1,1 then respond (internally) *yes* (that is, no problem); if the inputs are 1,0 (acoustic stimulation but no SR) then respond *no* (there is a problem). The child's simplest overt response to the *no* condition is to look puzzled or to say "Eh?" or "What?" This limited system will, of course, only deal with the case where zero semantic information emerges from the compiler.

To accomplish this simple function the inputs to the comparator need not really be ordered (or labelled). But eventually the comparator must be sensitive to order (or labelling) if the child is to gain access to the location and not merely the existence of input failures. We can illustrate the distinction in the following pair of exchanges. In the first conversation the child is 4;7, and the exchange is taking place in a noisy kitchen; the radio is on, and dishes are being washed. The topic--which has already been running for a couple of minutes--concerns what the child had been doing at school that day:

Adult: Did you enjoy yourself?
Child: What?
Adult: Did you have a good time at . . .
Child: No, no, *say* it again!

The adult has misunderstood the force of the "What?", thinking that the child had failed to understand when in fact she had failed to *hear*.

A month later, the reverse misunderstanding occurred. A typical dinner-time conversation is in progress:

Adult: And then the grape juice ferments and alcohol results from this pro-
       cess . . .
Child: What?
Adult: And the grape juice . . . (repeating the utterance with an identical
       intonation contour)

232

Child: No, no . . . in a *different* way so I can understand.

Again, the adult has misunderstood the point of the "What?", thinking this time that the child had simply not heard. The misunderstanding is soon corrected when the child shows how badly formulated (rather than quietly expressed) the adult's purported explanation was.

In order to express the distinction that the child is making we seem to require a system of the following nature:

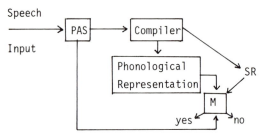

But now the decision rule for the comparator must become more complex. The input 1, 0, 0 (i.e., acoustic input, but no phonological or semantic representation) triggers the request to speak up (exchange 1), but the input 1, 1, 0 (i.e., acoustic input, phonological representation, but no semantic representation) triggers the request to express the *content* more clearly (exchange 2).

We have spoken so far as if failure to compile a semantic representation was an all-or-none process. But clearly this is not so. Consider the following exchange with a child of 4;9:

Adult:  You look very elegant in your new dress.
Child:  What does *elegant* mean?

Here the monitoring device which caught the (partial) failure to compile has been able to "dig out" and specify the linguistic unit which has caused the failure. It would seem that this requires an explanation at least as complicated as the following: The compiler produces a complete phonological code (no null symbols) but only a partial semantic code, that is a code with a hole in it (X1 X2 ..... $\emptyset$ ..... Xn). This null symbol triggers the routing of the available semantic code to the expressor which in turn produces a phonological code. We then need a new form of comparator, one which can substract one phonological code from another. The residue of this operation is that segment for which the compiler failed to find a semantic representation. This residue can then be used to initiate requests of the type "What does *residue* mean?" It is in this way that the child is "driving" her own acquisition process. Clearly, one can eventually learn what *elegant* means without asking. But it must be more efficient to ask rather than wait for the next twenty occurrences and hope to form the relevant inductive generalization on the basis of "context of situation." EMMA is thus a crucial part of the (internal) educational system.

Yet another form of monitoring operation would seem to be involved when the semantic representation is checked, not for holes, but rather for congruence with the context of situation in which the input utterance was produced. For example, a long telephone conversation with a child of 5;3 is well underway when the following exchange occurs:

Adult     :  What have you been doing at school today?
Telephone:  Click.
Adult     :  Hello?
Child     :  Why did you say *Hello*?

Clearly, the child's question has located the source (in this case, pragmatic) of the difficulty: Why say *Hello* when I've already been talking with you for five minutes? We presume then that (compiled) semantic representations are constantly monitored for their consistency with a much more general (and constantly updated) cognitive representation of the state of the world.

*Output Monitoring*

We now turn to the converse problem of specifying the ways in which the child can monitor and hence regulate and correct his own speech output. If we look at Clark's paper in this volume we see that in the earliest stages of language acquisition the child's mother serves as an external monitor. The following exchange from Scollon (1976) shows how lack of comprehension, real or feigned, on the part of the mother leads to a sequence of attempts to express the same content by the child, attempts which differ from each other by a small number of features:

Brenda:  [ š ]     (holding up mother's shoe)
         [ šI]
         [ š ]
         [šIš]
         [ šu]
         [šu?]

Brenda: [šuš]
Mother: Shoes! (p. 150).

These attempts seem to approximate more and more closely the correct form
and it is tempting (Scollon found) to conclude that the child is actually
listening to her own output and comparing it with an adult "standard." The
difference is then used as a means of correcting the next attempt. While
this is a possible substructure for the sequence we feel it is excessively
adult-omorphic. A simpler account might claim that the child's output representations are both incomplete (with respect to the adult "model") and,
perhaps, rather loosely "associated" with whatever semantic structures the
very young child is in command of. We have no intention of discussing here
the possible growth and learning mechanisms that are required. (In any revealing account it would be necessary to formulate these mechanisms at a
quasi-physiological level anyway.) A plausible set of theories would have
in common the feature that successive attempts to "activate" output structures become more and more successful. There seems to be no reason, on a
psychological level, to invoke a more complex mechanism than that which is
responsible for the phenomenon of operant conditioning (Marshall, 1970).

All that is necessary then is that the mother provide a simple error
signal. This might be the withholding of approval or simply silence. In
this situation the formulator is driven by a fundamental communicative rule:
"Keep talking until you get feedback (approval)." Thus the formulator keeps
sending the same semantic representation to the expressor. This SR is, in
effect, the address of the phonological code. The details of such sequences
may thus tell us about the way in which phonological codes are stored and
develop but nothing at all about complex monitoring. Examples in which an
*internal* monitor appears to be required, however, can also be found. For
example, the spontaneous correction, "I've been swin—swimming." Without
going into detail we only wish to observe that it is not necessary to think
in terms of the child listening to her own output and correcting it on the
basis of comparison with standard *adult* inputs. In a future paper we will
show how such errors are related to redundancy rules in phonological output
and to incompetence rules (Smith, 1973).

The previous example of error-correction involves comparison of articulatory or phonological signals. Clark's paper (this volume) also contains
cases where it would seem that the detection of mismatch must be in a semantic code. We shall illustrate with an example from another child. At age

2;1 the subject learned the word *caterpillar*; at 2;2 she learned *helicopter*.
Approximately two days after acquiring this latter word the following utterance was observed as a helicopter flew overhead: "Look, look! A caterpillar
. . . helicopter (laugh)." Since *caterpillar* is a good English word (and
was known to the subject at the time of the slip) we presume that the error
must have been detected by a mechanism of the following type:

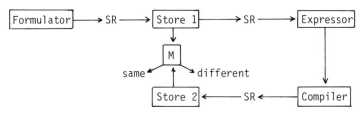

In light of the previous examples of phonological error, the nature of this
error is quite interesting. Although the error must be *corrected* by semantic
monitoring it is structurally closer to a phonological error. One explanation might run as follows: Given that *helicopter* was not well-entrenched in
the child's productive vocabulary, the phonological code available at a
particular moment may have been grossly underspecified, e.g., four segments,
the terminal one being /ə/. Clearly, such a representation is insufficient
to activate an articulatory event. On the assumption that *one* access code
for the dictionary is phonological, however, we could imagine that this
specification is sent to the dictionary as an instruction to find an item
which matches the code. *Caterpillar* is then the first word found which does so.

Let us now turn to morphological errors. Consider the following exchange with a child of 4;11:

Child:  I brang it home from school.
Adult:  What?
Child:  I bringed it home.
Adult:  Eh?
Child:  I brung it home.
Adult:  Oy vay!
Child:  Brought!
Adult:  What d'you know—we finally made it!

We seem to see here a complex interplay between two monitors, one external (a father) and one internal. The external feedback is minimally informative; it indicates that something has gone wrong but contains no cue to

the location or type of fault. Despite this vagueness of external feedback the child instantly homes in on the inflectional morphology of the verb. We might hypothesize then that an internal monitor is capable of assessing the state of the rules governing past tense inflection; specifically, that the monitor can provide a "confidence rating" for the pairing of stems with regular and irregular endings. A weaker interpretation, however, would claim that no monitor is required and that the sequence should be regarded as the recirculation of the original content through a system of unstable rules. The trigger for such recirculation could obviously be provided by the adult's (uninformative) error-signal. This latter interpretation may run into difficulties in explaining the child's deletion of content (initially *from school*, and eventually everything but the verb) from the hypothesized recirculation. Yet in other cases the weak interpretation may suffice. For example, a correction such as "It eat—it eats bread" does not even need the comparator. If we assume that the compiler can take as input the child's own output, then it would note the error in the output string which has violated the rule: Verbs in the present tense take /s/ when governed by a third person singular noun phrase. The compiler could then just feed back its compiled version to the expressor for re-encoding. This assumes that the error was due to noise in the system and not due to a systematic mischaracterization of the language in the expressor. In the latter case, of course, the result of the re-encoding would be identical with the original output.

Note that in this case no component need "know" which rule had been violated; the information need not be recoverable or available to another subsystem. All that happens is that one part of the normal language system sends a code to another.

A slightly more complex account of the error correction process leads to some interesting predictions. Imagine that the compiler can send back to the expressor not only the compiled version of an incorrect string but also a copy of the rule which has been violated. This is now a more complicated model because the form of the two pieces of information is different. Such "rule transmission" is clearly required if the perceptual system is to teach the production system. Assume that the compiler learns rules by examining the input strings. Then, once the agreement rule described above, or the regular past tense rule is induced, it could be transmitted from compiler to expressor in the way indicated. At an early stage in language acquisition the child will already be using some strong forms correctly (*went, ran, was,* and so on) and perhaps some correctly, but non-productively, inflected weak forms. As soon as the compiler acquires the regular rule the strong forms will be changed to correspond to the rule-governed form. This fits with the facts of acquisition and also predicts that children will "correct" their own strong forms to the incorrect weak form. One such correction has been reported by Bever (1975):

Child : Mommy goed to the store.
Father: Mommy goed to the store?
Child : No, Daddy; I say it that way, not *you*!
Father: Mommy wented to the store?
Child : No!
Father: Mommy went to the store.
Child : That's right, Mommy wen . . . Mommy goed to the store.

Faced with examples like the above, one is tempted to move immediately to a "high level," "abstract" description which makes specific reference to awareness, consciousness etc. And indeed the child *does* appear to be aware that his own usage and that of his father differ. Similarly, in the following examples (from a child aged 5;3), the child gives every indication of being fully conscious of what has gone wrong; the child's introspections are of an adult character.

Adult:  (to second adult) Are you going to put the garage in the car?
Child:  Ha, ha, ha. Daddy got it wrong again.
Adult:  What should I have said?
Child:  Are you going to put the car in the garage.

While we have no objections to using a terminology that includes conscious and self-conscious processes, the danger is that such descriptions can easily lead us to lose sight of the need to formulate a mechanism whereby the child (or adult) has *access* to the subparts of structural descriptions involved in primary language processes.

*Conclusions*

   We believe, then, that "awareness" arises from the operation of error-detecting mechanisms which have access to subparts of the output of primary production and comprehension systems. It is, of course, the sheer complexity of linguistic programs (and the conditions of their appropriate use) that requires the development of fault-finders and fault-describers. Failure to

understand can take place at any level between (and including) failure to
hear the physical token and failure to grasp why on earth a token of that
type should have been uttered under such-and-such circumstances. Failures
to express an intelligible signal can arise on a similarly wide range of
levels from conceptual confusion to sloppy articulation. We have accordingly tried to outline some of the mechanisms which are required to change
errors into correct responses.

However, given that the relationship between form and content (i.e.,
phonological and semantic representation) is *conventional* we could just as
well use the same devices to change correct utterances into erroneous ones.
It would seem that the child, by reflecting upon his own linguistic skills
and knowledge, does indeed eventually become aware of the fact that linguistic behavior is rule-governed and not law-governed. The best known discussion of the development of such awareness is, of course, to be found in
Vygotsky's claim that the young child initially believes the name of an object to be an intrinsic *part* of that object and only later realizes that you
can call it anything you like (although it helps to get other people to do
likewise). Although it appears that Vygotsky (1962) underestimated the
sophistication of even very young children, the issue that he drew attention
to is an important one. The ability of the child to abstract himself from
the normal use of language enables him to play games with the relationship
between sound and meaning. Thus at age 4;11 one child invented a nerve-
shattering song that consisted simply of repeating ad nauseam (albeit at
ever increasing intensity) the line: "Yes is no and no is yes." The same
child at 5;1 announced that "If I say I'm not tired that means I am tired,
and if I say I'm tired that means I'm not." For the next fifteen minutes
(before getting bored with the game) she proceeded to replace the content
words in her sentences with their antonyms (or other words drawn from the
same semantic field) at every available opportunity. For example, she lay
on the floor and said "I'm standing on my bed," and sat at the kitchen table
at six p.m. saying "Where is my breakfast?" Young children demonstrate quite
explicit awareness of the conventional nature of language. For example, one
child at age 3;1 remarked, "Mummy says *mato* and Daddy says *marto*." Slobin's
paper (this volume) contains other illustrations.

These last examples are clearly ones for which everyone will be happy
to implicate EMMA (in the above sense of explicit awareness). But what of
the dividing line? We may now confess to having led the reader along the

false trail we ourselves followed. As soon as one postulates any kind of monitoring device, even the very simplest comparator, we see an embryonic EMMA. Thus, rather than being pushed out to the mists to which awareness has been consigned, EMMA functions can be seen at a very early age, and rather than being more complex than Normal Language Processes EMMA appears actually to be simpler. One might accordingly decide to restrict language awareness to EMMA plus consciousness. But that, of course, is another story.

Finally, we would hope to find ontogenetic continuity in EMMA function; just as individual children evolve production strategies in different ways we would expect to find differences in the way their monitoring devices evolve. Our attempted demystification of the developing processes of correction is then also an explication of EMMA and our eponymous heroine has a name which is already an anachronym: For she seems to be rather an Ever More Maturing Adjunct.

*ACKNOWLEDGEMENT*

We thank Wolfgang Klein and William Marslen-Wilson for a number of helpful suggestions. Jerry Bruner encouraged us to try to find a positive role for awareness. And Zoë L. Marshall provided many insightful comments on all aspects of language-learning.

*REFERENCES*

Bever, T.G. Psychologically real grammar emerges because of its role in language acquisition. In: D.P. Dato (ed.), *Georgetown University Monograph Series on Languages and Linguistics*. Washington, D.C.: Georgetown University Press, 1975.

Brown, R., & Bellugi, U. Three processes in the child's acquisition of syntax. *Harvard Educational Review*, 1964, 34, 133-151.

Chomsky, N. *Aspects of the theory of syntax*. Cambridge, Mass.: M.I.T. Press, 1965.

Chomsky, N. On the biological basis of language capacities. In: R.W. Rieber (ed.), *The neuropsychology of language*. New York: Plenum Press, 1976.

Crowder, R.G., & Morton, J. Precategorial acoustic storage (PAS). *Perception and Psychophysics*, 1969, 6, 365-371.

James, W. *The principles of psychology*. London: MacMillan, 1890.

Marshall, J.C. Can humans talk? In: J. Morton (ed.), *Biological and social factors in psycholinguistics*. London: Logos Press, 1970.

Scollon, R. *Conversations with a one-year old: A case study of the developmental foundation of syntax*. Honolulu: The University Press of Hawaii, 1976.

Smith, N.V. *The acquisition of phonology: A case study*. Cambridge: Cambridge University Press, 1973.

Vygotsky, L.S. *Thought and language*. Cambridge, Mass.: M.I.T. Press, 1962.

# Appendix

## The Role of Dialogue in Language Acquisition*

J.S. Bruner

Wolfson College, University of Oxford, Oxford, Great Britain

I am profoundly moved by this occasion--the establishment of the Max-Planck Project Group in Psycholinguistics. It is not only a signal recognition of the growth of the psychology of language, a field that has made extraordinary progress, but also represents a much needed focusing of effort upon an enormously complicated subject. As our wise colleague Roger Brown has put it, speaking of the relations between psychology and linguistics in the study of language acquisition, "We psychologists have no choice but to enter into the enterprise of linguistics, and linguists, if they think about it, have no choice but to pay attention to work in developmental psycholinguistics. The study of language is one, partialled out among academic departments to accomodate the limitations of human intelligence but, ultimately, one subject. The sciences do not build like pyramids with the higher levels waiting for completion of the lower. They build like partial views of Leviathan growing toward one another and a complete view." Let this project group and whatever it brings into being in the years to come serve then as a battery of lenses where we can look each in our own way at the myriad facets of this most intricate human enterprise, language, and share what we see with each other. Some years ago, I bought an 18th Century print in Milan of a Planisferio Celeste in the four corners of which are proudly represented the then reigning observatories of Paris, Kassel, Greenwich, and Copenhagen. Mapping language will be more difficult than mapping the heavens. If ever our print makers are inspired to celebrate it with maps, I hope that Nijmegen will be a figure in one of its corners. I am presenting this Planisferio to the Group as a token of Opening Day.

---

* (Inaugural Lecture of the Max-Planck-Gesellschaft, Projektgruppe für Psycholinguistik)

And now let us return to the serious business at hand--the manner in which dialogue affects the acquisition of language. Because we are a mixed audience, representing many disciplines other than linguistics and psychology or some intersect of the two, I must begin with some preliminary remarks on why so obvious a topic, one that might seem even banal to the layman, should in this year of founding be considered fresh and lively and at the growing edge of knowledge.

The acquisition of language has always been a puzzling phenomenon. How can human beings learn so complex a system of rules for producing and comprehending messages so quickly, so well, with such subtle flexibility of use, a system of rules so complex that we who concern ourselves with language can scarcely decide how to describe it? Why did not psychology provide the answer forthwith? Indeed, we can now see that psychology may have added to the difficulty. For psychology has from its start suffered from a built-in difficulty in dealing with the acquisition of knowledge, knowledge of the kind implied by "knowing how to speak." Psychologists were philosophically committed to a reductionism that produced the difficulties. One aspect of the reductionism was the assumption that "complex" learning was in principle reducible to the concatenation of "simpler" forms of learning--an associative concatenation. By the same token, "knowledge" of a language was thought to be reducible to a knowledge of a simpler kind, the more complex form being produced by a concatenation of the simpler elements. Since knowledge of a language and its structure does not reduce in that concatenative way, the psychologist's approach was bound to beg for trouble. A second feature of the reductionism of "learning" theory might well be called the "postulate of initial innocence." Properly to study the acquisition of knowledge, according to this postulate, one must begin with an organism who knows nothing about the domain and track the process through the stages by which "possession" of the knowledge is finally achieved. This imposed the requirement that the knowledge to be acquired had to be brand new and be of a kind, moreover, such that its gradual acquisition could be studied. One could not, obviously, study the acquisition of meaningful knowledge, for example sentences like "The fearless rat jumped over the lazy cat," even though such a sentence was fresh and new knowledge, because it takes no time at all to learn it, one trial at most, and because, it was said, such sentences could not be described without references to previous associations and previous knowledge. One wanted to study learning *in vitro* and *ab initio*. There were sceptics like William James who knew well what later I shall refer to as the Apostel principle: that to learn something about a domain requires that you already know something about the domain and that, perhaps, there is no such thing as *ab initio* learning pure and simple. So these good men and women developed paradigms for the study of learning that involved the learning of nonsense syllables and unconnected word lists and finger mazes. To be historically accurate, there were non-conformists among them who urged that inherent structure and meaning and rule sensitivity were crucial characteristics of all knowledge acquisition and that the nonsense materials when used as "stimuli" in experiments on learning falsified the picture--and we celebrate such masters as Köhler, Bartlett, and Tolman for standing out against the prevailing anti-intellectualism of what may still be called "modern" learning theory. The principles that have come out of such modern learning theory were mainly associationistic in character, not terribly different in emphasis from the associationism of Aristotle, of the British empiricists, of Herbart. To this simple concept of association by contiguity and similarity was added a doctrine of reinforcement to accommodate the new behaviorism that had come to prevail after the first World War. For by the new dispensation, when one learned what things went together, one was learning to make a response to a stimulus and that response could be acquired by some form of reward or by achieving some form of instrumental end.

There was also another persistent fiction in the early efforts to investigate the acquisition of knowledge. Not only was there an implicit insistence that the learning be *ab initio* and that it be autonomous and therefore *in vitro*, but that it be a form of learning that was, once achieved, unambiguously "in possession" of the learner. This is a serious bias. It stemmed from our use of the criterion of "perfect performance," which meant for the subject being able to prove to the experimenter that one had *full possession* of the *particular knowledge* in question, and one could show that most unambiguously by *repeating* what had been learned. This insistence--stemming from methodological rather than philosophical considerations in the main--is at variance with at least two characteristics of the acquisition of organized knowledge. One is in the very nature of *generativeness*: that in learning anything, this permits us to go beyond that mastery to generate new forms of knowledge, or at least extensions of what we have learned. The apprentice shipwright at the famous Abeking and Rasmussen yacht yard was, at the end of his period, required to build a faultless yacht-tender from a set of plans, though never given any explicit training in doing so. We take it for granted that, in much the same spirit, language has been learned when one can say something one has heard in one's own way, in paraphrase. But the other difficulty with the learning theorist's paradigm was not only that "possession" was measured in an irrelevant way as far as language *competence* was concerned, but that it failed equally to take into account the inherently social nature of what is learned when one learns language and, by the same token, to consider the essentially social way in which the acquisition of knowledge of language must occur. We use language in order to communicate with others, and as John Searle somewhere remarks, one could possess a quite rich and full knowledge of syntax and semantics and still be rather a hopeless idiot as far as communicative competence is concerned. Communicative competence has to do with dialogue. But before I take up that matter, I should like to turn now to the dilemma of the linguist in applying the psychology of learning to the phenomena of language acquisition. It is, alas, a bit like C.K. Ogden's parable of the man who wished to find out about Chinese metaphysics and consulted first "China" and then "Metaphysics" in the *Encyclopaedia Brittanica*.

For when the cumbersome apparatus of explanation-by-learning-theory came to be applied in detail to the acquisition of real language (in fact, it had been applied metaphorically some time before by no less a figure than St. Augustine), the picture that emerged was a very dim one indeed. Our distinguished colleague George Miller once computed the number of correct sentences in length from two words to fifty that an ordinary mortal was capable of producing or comprehending and the number of reinforcements that would have been required to generate and establish that set in contrastive discrimination with all grammatical strings, and concluded that it would be of the order of $10^3$ per second for the length of an entire lifetime. His assumptions may have been little shaky in view of the escape hatches through which conventional learning theorists usually slip, so we might reduce the figure to order of magnitude $10^2$ per second!

It will come as no surprise, then, that when Noam Chomsky came on the scene in the 1960s and, once for all, dismissed the "classical" or associative conception of language learning and replaced it with one that freed the developmental psycholinguist, again seemingly, from the burden of considering learning at all, a certain number of us were prepared to welcome him as a new liberator. Though what he had to say about language acquisition now appears to be grossly wrong, as I shall try to relate, it was *productively* wrong and enormously fruitful. It went something like this--and I cannot hope to match the perceptive essay by Professor Levelt (1975) entitled "What became of LAD?" (Chomsky' Language Acquisition Device). Language is not learned in any conventional sense. Innately, the child possesses a knowledge of language universals and, merely as a spectator at the feast of adult language, he recognizes the universal principles that are realized in the local language to which he is exposed. There is nothing privileged about his linguistic interaction with adult vicars of the linguistic community, nor does he need real-world knowledge of the domains to which the language refers. The first acquisition is of the syntax of the language. Language and its rules are a domain separate from anything else, autonomous.

What was productive about this view--and let us not totally dismiss its message quite yet--was that in a stroke it freed a generation of psycholin-guists from the paralyzing dogma of the association-imitation-reinforcement paradigm. It put in its place, momentarily but crucially, a model that pro-posed that the child at the start of language learning grasped rules that made it possible for him to generate sentences that were grammatical and none that were not (or few, and those for trivial, performance reasons), and that these sentences were generated from a deep structure via transformation rules that created the surface level of spoken language. Part of its pro-ductivity was its extremeness. It was bound to produce counteractions. LAD as a strong doctrine never made it to its tenth birthday.

The counteractions that have occurred since the formulation of this brave doctrine in the 1960s have been several in number, and I want to sketch these briefly before turning my attention to the last of them. I should re-mind you that I am speaking as a psychologist rather than as a linguist, for the implications of what I have to recount are somewhat different for the two tribes. The first counteraction was about real world knowledge. In fact, real world knowledge is important for the mastery of language. Pre-linguistic concepts provide guides for the learning of forms of utterance that relate to them or refer to them. The work of Sinclair de Zwart (1967), of Roger Brown (1977), of Katherine Nelson (in press), and more recently of Rosch (1973) and of Anglin (1977) argues overwhelmingly that the child sorts out his universe conceptually into categories and classes, is able to make dis-tinctions about actions and agents and objects before he has the language for making those distinctions in speech. "The concept is ... there before-

hand, waiting for the word to come along that names it" (Brown, 1977).
It still remains a mystery how the child penetrates the communicative system
and learns how to represent in language what he already knows in the real
world--i.e., knows conceptually. For what he is learning about language
(though it already has semantic content) is not the same as what he knows
about the world. Yet it is a big step ahead to recognize that, at the start,
the child is not flying blind, that semantically speaking he has some target
toward which his language-learning efforts are directed: saying something
or understanding something about events in the world that he already knows.

The next step (and you will forgive my giving the impression of a chronology, for in fact the chronology is more in the imagination than in history)--the next step was a functionalist one. That is to say, if the child
were in fact communicating about a matter of which he already had some knowledge, he presumably was doing so with some end in mind, some function to be
fulfilled. He must be requesting or demanding or referring or establishing
some sort of interpersonal relationship. Now, there is a long history of
linguistic functionalism and many linguists have made a bit of the fact that
there is something like an expressive function or a referential function or
a conative function in speaking. Because nothing much had come of this insistence, the functional field had been allowed to lie fallow for some years.
Two new developments brought it back into lively cultivation. One was a
revolution within linguistic philosophy and the other within what might more
properly be called socio-linguistics. The first, of course, was sparked by
John Austin's brilliant *How to do things with words*, his William James Lec-
tures at Harvard in 1955, though the impact of his thought was not felt in
linguistics until over a decade later. His argument was--or is, in the form
of its extension by Grice, Searle, Strawson, and others--that an utterance
cannot be analyzed out of the context of its use and this must include the
intention of the speaker and the interpretation of that intention by an
interlocutor. Any utterance is a speech act that can be analyzed into a
locution, or its content and structure; its illocutionary force, or the conventional intent embodied in a linguistic procedure; and its perlocutionary
force, or the inadvertent consequences that it produces. A speaker may use a
conventional interrogative construction for making a request, do so by making
a declarative statement, or use the imperative--so long as each represents
a convention of the linguistic community that assures its "uptake." Roger
Brown relates an interesting point: in the protocols of Sarah, he found that

her middle-class mother used the convention of making requests by the use of interrogatives: "Why don't you play with your doll now?" Once Sarah had come to appreciate what I shall call "genuine" *why* questions, "Why are you playing with your doll?," she typically answered these with the well-known "Because." But there is no instance either before or after the appearance of the comprehension of the causal question of her ever confusing a sham and a real *why* question. She evidently recognized the illocutionary force of the two forms of utterance quite adequately. Indeed, Smith and Greenfield (1976) in their recent book argue that the earliest holophrases of the child are pure performatives, where the illocutionary force, so to speak, is the exclusive content, the child's early "Mummy" being the same as when an adult says "I promise." I think this is a mistaken interpretation of what Austin had in mind by a pure performative, but it illustrates the push now in progress.

It was perhaps Halliday (1975) who gave the biggest impetus to the reintroduction of functional categories into studies of language acquisition in his *Learning how to mean*. His argument parallels the semantic one: not only does pre-linguistic conceptual knowledge precede linguistic progress, but so too does function precede it. The child knows (in limited form) what he is trying to accomplish before he begins to use language to implement his efforts. He tries initially by gesture and vocalization and, at least in the case of Nigel, Halliday's son and subject, to signal the function in characteristic ways--as in a rising intonation pattern for pragmatic functions and a falling intonation for informational or "mathetic" ones. Halliday sets forth a list of the functions and proposes hypotheses about the order of the achievement. What is challenging about his formulation is that he argues that language proper, lexico-grammatical speech, is only one of the procedures by which the child operates, but that it has the powerful property, unlike gestural and vocal ones, of making possible the simultaneous fulfillment of all his seven functions, from the simple regulatory one to the imaginative one. And for him, some of the higher order functions (the imaginative and the informative) require the constructional use of language to be fulfilled at all. In short, then, if the semantic and the functional primacists are to be taken seriously, language emerges as a procedural acquisition to deal with events that the child already understands conceptually and to achieve communicative objectives that the child, at least partially, can already realize by other means.

But as you can see, this still leaves as a mystery how it is that the child masters the complex rules of language, language proper. Were I reporting on my own research on the transition from prelinguistic to linguistic communication, I would want to urge upon you that it is just as mysterious how the child adopts *any* "non-natural" or conventionalized mode of communication, that *language* acquisition has no monopoly on the dark mantle of mystery. But language is the end result of a long series of developments en route, and is properly where our attention should focus. And now, finally, I must turn to the role of dialogue. For language is the medium of dialogue, and it is in dialogue, that knowledge of language *per se* develops.

In looking to dialogue as the source of the child's learning of the various aspects of language--syntactic, semantic, and pragmatic--there are several ways to proceed, and all are being hotly pursued. It is still a little difficult to achieve an overview of the bustle, but before we plunge in, perhaps it would be worth a try. A first approach is to examine the properties of "Motherese" or Baby Talk, or BT, the language of adults to which language-acquiring infants are exposed. This would at least provide us with a picture of the Input to any Language Acquisition Device that may be operative in the child and is, in a real sense, staying true to the original Chomsky LAD program. For in fact there is no reason to assume that a Language Acquisition Device should be ignored as a possibility--one as rich as is necessary. I shall call this an *Input Model*. A second approach looks to dialogue as providing constraints and conditions on the child's output, and it can be broadly conceived as an *Output Regulation Model* in which the child learns to achieve communicative results by reference to the effectiveness of his role in the dialogue with an adult. The third approach is a *Transactional Model* in which the central assumption is that the child is learning not so much to penetrate the rules of somebody's input or to regulate his own output, but rather to understand the rules of dialogue *per se*, the necessary presuppositional structures on which it is based, the principles of deixis whereby the context of a speaker-listener pair can be mapped into their exchange, and the ways in which intentions of the two partners can be both signalled and interpreted. It can be said of the three approaches that they are all based upon an underlying assumption that the language learner is an active hypothesis generator and that his hypotheses are derived from his recognition that there are communicative procedures to be mastered that will fulfill these functions. I rather suspect that the three approaches

are really one: that in the end the transactional approach will become sufficiently rich to subsume the others. But I want to keep them separated
for the moment in order to highlight some specific studies that seem to be
tending in somewhat divergent directions.

Consider the Input studies first, for in many ways they provide the
easiest access. There are some interesting, almost universal properties of
Motherese or BT, and Roger Brown provides us with a nice overview of these
in his Introduction to the recent Snow and Ferguson volume (1977), *Talking
to children: language input and acquisition*. The BT register appears according to Ferguson in that volume to have three kinds of processes that generate
its features: *simplifying*, as in replacing difficult consonants with easy
ones or replacing deixis-demanding pronouns with proper names; *clarifying*,
as in speaking slowly, repetitively, and with exaggerated segmentation; and
*expressive*, as in using nursery tone, higher pitch, hypocoristic affixes and
baby-like euphemisms, and so on. Brown suggests that these might well be
collapsed to two processes: a *simplifying-clarifying* one that has the function
of getting communication established or assured and possibly of "teaching"
the child, and an *expressive* one that has as its major aim to express affection and/or to capture the child's attention. He proposes that the two generate features that have to do respectively with coping with communicative incompetence and with inspiring affection or attention and that BT can be
thought of as a "complex clot in the linguistic stream" which can be clotted
in different ways as used for other registers--as when we talk with emphasis
on affection-achieving features to animals or to dice we are casting, or with
emphasis on communication-facilitating features to foreigners or to adult
retardates or to people suffering the shock of injury.

This then leaves us with some questions as to why it is that we are
so conscious of Baby Talk and why we should assume that it has any role in
helping the child learn the language. We shall leave out of account the
origins of BT, a matter that intrigues Brown and various of the authors in
the volume mentioned, save in one crucial regard. The one speculation about
BT that must concern us in trying to answer our question about its role in
teaching language is the hypothesis that BT may be based on an adult's model
of TB--that the Baby Talk of adults may be based upon adult transformations
of Talking Baby as practiced by infants. Adult BT and the speech of babies
both share a higher fundamental frequency than speech between adults, both
substitute proper names for pronouns, both are high in imitation and repeti-

Register
Anglophone.

tion; both show low upper bound on length, reduced semantic complexity, limited semantic relationships in multi-morphemic utterances. And so on.

But as Brown notes, does this not create an extraordinary paradox? As he remarks, "Babies already talk like babies, so what is the earthly use of parents doing the same? Surely it is the parent's job to teach the adult language." He answers this in a form with which I am fully in agreement, although for somewhat different reasons than his. "What I think adults are trying to do when they use BT with children, is to communicate, to understand and to be understood, to keep two minds focussed on the same topic." He is undoubtedly right, but later I shall argue that there is more to it than that. What impresses him and moves him to his conclusion are some striking results in studies by Cross in the same volume. She correlates two sets of measures, one taken from the mother, the other from the infant. The latter is based on the mother's success in being able to interpolate 100 syntactically diverse sentences in small doses at natural junctures in her speech to her child. Each sentence was judged for its comprehension, verbally or behaviorally, thus yielding a "receptiveness score" for the infant. She also counted all manner of BT indicators in the mother's speech: imitations, expansions, repetitions, higher pitch, etc. What cannot help but impress one is the striking correlation between what the mother does and what the child is doing: the mother's performance is obviously highly-tuned to the child's level of performance. To take the two examples that most impress Brown (and me as well), there is a -.85 correlation between receptiveness and the use of expansions in the mother's speech: the less receptive the child, the more the mother checks by expanding on what the child has said as if to confirm. And the correlation between receptiveness and mother's reference to non-immediate events is +.72. I would add to this finding another more direct measure based on a study by Greenbaum and Landau (1973), who found in their Israeli and Bedouin infants, studied through a full day, that forms of discourse used by the mothers to their infants tend to disappear to a minimal point the moment the child begins to be able to respond to them. At that point they are used purposefully, rather than as part of a general BT register. A notable example is the use of the interrogative whose use plunges down at the point where the infant begins responding to questions. Then, questions are used purposefully with the intent of eliciting an appropriate answer. We have found the same phenomenon in our own work, but I shall return to the issue later. Indeed, Brown makes the same point succinctly: "BT is elicited

Mother
tuned
into
child's
level.

by indications of some psycholinguistic ability rather than by ... age."

This brings us to question what is the significance of the mother's BT input looked at just as a *sample* of speech to which the infant is exposed--in contrast to the view of a decade ago that it made not much difference how degenerate a version of speech the child was exposed to, given his innate capacity to extract local realization rules from any old input in the language. If the mother's BT is only imitating what the child can already do, *1.* what indeed is the function served? And here we arrive at either the output regulation model (which would argue, weakly, that the mother's repetition of the child's speech is in the form of a confirmation of his hypotheses), or *2.* the transactional model. The transactional model would say of what we have considered thus far something as follows. The mother indeed uses BT to communicate and, in an actuarial sense, to provide the child with a model that *3.* culturally speaking has been assumed to be easy of access to the child--in terms of clarity, simplicity, attentional vividness, and affectional attractiveness. She does it to get communication started (a matter to which we shall return), but the moment that she gets communication started, the main task of teaching the uses of the language begins. And here I think it is not without consequences that Roger Brown and I have spent nearly twenty years as colleagues, for I can imagine nobody saying more exactly what I now intend to say, and I shall introduce it in his words: "A study of detailed mother/child interaction, and I have done my share of it, shows that successful communication on one level is always the launching platform for attempts *4.* at communication on a more adult level." This is the Apostel principle again --again, more in a moment. Since I have derived such riches from Brown's work over the years--and from his speculative ventures as well--let me attempt now to repay him a little by an account of some of the ways in which leaps forward in language use are accomplished not only by effective deployment of already mutually understood Baby Talk, but also how the mother systematically changes her BT in order to "raise the ante" or alter the conditions she imposes on the child's speech in different settings. *5.*

I begin with a point that is familiar to all. Mothers invariably have a theory of the child's linguistic ability and performance. The first premise of all such theories is that the child is trying to communicate that, in some biologically authentic way, he is an agent capable of intending to *6.* achieve something interpersonal. That is one of the reasons why mothers so frequently experience such pleasure when a child can first sustain eye-to-eye contact and why they are so disturbed when he cannot (Robson, 1967)--why they so welcome all signs of reciprocity. Indeed, there is evidence to indicate that child abuse is sometimes associated with non-responsiveness in the child. The second point that needs making is that the range of situations in which mother and infant find themselves during the infant's first year or so are highly limited in number and kind. They include such familiar formats as activating attention to objects, give-and-take exchanges, peekaboo and object appearance-disappearance-reappearance routines, feeding, various forms of caretaking which though routinized are also varied for fun and for the dissipation of boredom on both sides. In time and with variation, these formats are internally highly differentiated conceptually and become marked by more or less idiosyncratic ways of communicating by non-linguistic means. The functions of the communication that is going on are not highly varied at first, but become increasingly so as the first year wears on. There is a great deal of requesting which in time becomes differentiated into requests for different objects, different agents, different actions, different possession relations, different locations. An enormous amount of the communication is also given over to the management of joint attention: achieving a common attentional focus and then achieving some elaboration upon the focus that begins increasingly toward the end of the year to be in the form of joint topic-comment structuring. And much communication is also related to maintaining and elaborating an interpersonal link.

It is instructive to trace the formats that mother and infant share and examine their developmental course. I would like to consider briefly one of them: "book reading" with a baby. The point at which we enter is already well along: the child already knows a great deal about it, in the Apostel manner, and is not starting *ab initio* either procedurally or conceptually. It is a well-known format in which the mother shows the child pictures in a book, the manifest point of the exercise being to teach the child the names of the objects indicated. It is from a study carried out jointly with Anat Ninio and only recently published (Ninio & Bruner, 1978). The very appearance of the format is contingent on two prior developments and it represents exactly what Roger Brown means by taking off from a prior launching platform. For before "book reading" starts, the child has shown two other forms of communicative ability. One is the use of pointing as a means of ostensive referring--a long and slow process of development (Bruner, 1975). The child has shown the mother that he can point to objects that catch his attention.

look back to the mother, and utter some form of demonstrative that is a variant of "Da." His second accomplishment is evidenced by what Dore (in press) refers to as PCF's--phonetically consistent forms showing four properties: they are readily isolable and bounded by pauses, they occur repeatedly as items in a repertoire, they are more phonetically stable than babbling but less so than words, and can be loosely correlated with aspects of the environment or with aspects of an action. Once these two accomplishments have been assured, the mother begins in "book reading" seriousness.

Her first effort is to establish a new format: the book is to be looked at, not torn or eaten or thrown. Her own utterance pattern then begins to take on the usual steady pattern of fine-tuned contingency to which Cross refers. It consists of three major components and a forth optional terminal component. The first is an attentional vocative: *Oh look, Richard* as she points to a target picture, followed by a *Wh*... question, *What's that, Richard?* usually timed to coincide with his change in line of regard toward the picture, and, after a pause, a *Label*. The junctures are timed or phased to the child's response. If the initial attentional vocative does not take, it is repeated. If the child does not reply by some vocalization, the label is presented after a pause. If he does vocalize--and over several months he responds vocally with increasing frequency--the response of labelling is given in confirmational form, no matter how he vocalized, by some such utterance as "Yes, that's a rabbit." If the child vocalizes after the label, it is greeted with a confirming optional "Yes, that's right." In time, the child begins imitatively or otherwise, to substitute either shorter CV's and CCV's or VC's after the request for a label. Very shortly afterward, the mother raises the ante and will no longer accept an indifferent babble in reply. *What's that* now requires some approximation to a short burst. When PCF's appear, she will no longer accept other versions of vocalization.

If we examine the likelihood of the child giving a label or PCF in response to the mother's *Wh* question in contrast to imitating the label she has given in the penultimate position, we find the former exceed in the ratio of 4:1 over the latter. The child is recognizing the point at which a response is required, and the recognition is of an appropriate point of entry into the dialogue. In time, moreover, the mother gives over more and more of the initiation and pacing of the format to the child, and comes systematically to recognize by her utterances distinctions in the child's mastery. One indication, for example, is that the mother's initial attentional vocative drops out when the picture in question is one for which the child has a name: she now begins with a _Wh_ question delivered with a distinctive intonation contour--a falling terminal stress rather than a rising one.

*Foray into teaching*

In time, once the format has been well established and its variants worked out, the mother will use it as a "launching platform" for a next foray. But foray into what? Is she teaching? Well, yes, but if you should ask, she is also doing it because it amuses, is fun, keeps her child occupied and happy. The fact of the matter is that the only way in which the child can be kept communicating in a format is by altering or varying it--he or his mother taking the lead. In this case, the next step was moving from labelling into the establishment of topic-comment structures: "What's the pussy _doing_?" And here again, as we know from Howe's work (unpublished), the mother is likely to simplify the domain by concentrating upon a few conventional propositions of state or of action.

I have used the expression "scaffolding" to characterize what the mother provides on her side of the dyad in one of the regularized formats--she reduces the degrees of freedom with which the child has to cope, concentrates his attention into a manageable domain, and provides models of the expected dialogue from which he can extract selectively what he needs for filling his role in discourse. But she also does two other things as well. One of them is properly called "extension." It consists in her extending the situations in which and the functions for which different utterances or vocalizations can be used. I do not believe that, in the main, this is done on purpose, so to speak. I think it happens principally to keep the child's interest, to keep communication going. Virginia Woolf remarks in her _Writer's Diary_ at one point that she knows her novel, _Orlando_, is approaching its end, for she is getting bored with it. I sometimes think the managements of boredom plays a larger part in human affairs than we care to credit, and it may well be that linguistic extension owes something to it. Whatever the motive behind such "extension" its effect is to widen the range of contexts in which particular kinds of dialogue exchanges occur. It provides an ideal opportunity for the child to observe and master the different senses of words and expressions and the different uses to which they can be put.

Finally, the mother plays the role of communicative ratchet: once the child has made a step forward, she will not let him slide back. She assures that he go on with the next construction to develop a _next_ platform for his _next_ launch. My collaborator, Kathleen Sylva, and I have seen the process in operation repeatedly: insistence that the child pause in demand calling before she responds, once he *has* paused; insisting that he not grab once he has indicated an object of desire by other ostensive means, etc. She is a guardian of newly confirmed communicative hypotheses: a very crucial role.

Ladies and gentleman, I am fully aware that the account I have just given is not sufficient to account for the grasping of grammar or even for the mastery of semantic or pragmatic/illocutionary rules. I have wanted only to remark that however such mastery is attained, it has an enormous dialogic component. I think I can end by repeating the conclusion of Roger Brown's paper from which I have already derived so many ideas and which I should like to honor. He comments on the parental question so often asked those of us who work on child language and its development: How can we help the child to learn language? "Believe that your child can understand more than he or she can say, and seek, above all, to communicate. To understand and to be understood. To keep your minds fixed on the same target. In doing that, you will, without thinking about it, make 100 or 1000 alterations in your speech and action. Do not try to practice them as such. There is no set of rules of how to talk to a child that can even approach what you un-consciously know. If you concentrate on communicating, everything else will follow." To which I would add, the best practice for mastering dialogue is to enter into it. Give the child his chance. And at that, you don't have to give much of a chance, for everything we know about the period prior to the mastery of language proper indicates that the child has been hard at work mastering the subroutines that make dialogue possible--even in the first half of the first year when, according to the pioneering work of Stern (1978), and of Bateson (in press) the mother and infant are well launched in develop-ing a convention of turn-taking in gesture and vocalization. *Gesture + vocalization + turn taking*

In the end, indeed, it is very much like the mapping of Leviathan. There are many bits already well-discerned and, conceivably, a rough outline of the big and powerful beast is coming into view. It remains only for the right personage to declare this observatory officially opened, with the right per-formative! If that is my function, I duly exercise it:

INCIPTOTE QUO HOC FIAT FACILIUS
Start ye so that it be more easily done!

*REFERENCES*

Anglin, M. *Word, object and conceptual development*. New York: Norton, 1977.

Bateson, M.C. Mother-infant exchanges: The epigenesis of conversational interaction. In: R. Rieber & D. Aaronson (eds.), *Annals of the New York Academy of Sciences*, in press.

Brown, R. Introduction. In: C. Snow & C. Ferguson (eds.), *Talking to children: Language input and acquisition*. New York: Academic Press, 1977.

Bruner, J.S. From communication to language: A psychological perspective. *Cognition*, 1975, 3, 255-287.

Cross, T.G. Mothers' speech adjustments: The contribution of selected child listener variables. In: C. Snow & C. Ferguson (eds.), *Talking to children: Language input and acquisition*. New York: Academic Press, 1977.

Dore, J. Conditions on the acquisition of speech acts. In: I. Markova (ed.), *The social context of language*. New York: John Wiley and Sons, in press.

Greenbaum, C.W., & Landau, R. Mother's speech and the early development of vocal behavior: Findings from a cross-cultural observation study in Israel. In: *Cultural and social influences in infancy and early child hood* (Burg Wartenstein Symposium No. 57). New York: Wenner-Gren Foundation for Anthropological Research, 1973.

Greenfield, P.M., & Smith, J.H. *The structure of communication in early language development*. New York: Academic Press, 1976.

Halliday, M.A.K. *Learning how to mean*. London: Arnold, 1975.

Howe, C. The nature and origin of social class: Differences in the propositions expressed by young children, Doctoral Thesis, University of Cambridge, forthcoming.

Levelt, W.J.M. *What became of LAD?* The Hague: Peter de Ridder Press, 1975.

Nelson, K. *Learning to talk*. Cambridge: Harvard University Press, in press.

Ninio, A., & Bruner, J.S. The achievements and antecedents of labelling. *Journal of Child Language*, 1978, 5, 1-15.

Robson, K.S. The role of eye-to-eye contact in maternal-infant attachment, *Journal of Child Psychology and Psychiatry*, 8, 13-25, 1967.

Rosch, E.H. On the internal structure of perceptual and semantic categories. In: T.E. Moore (ed.), *Cognitive development and the acquisition of language*. New York: Academic Press, 1973.

Sinclair de Zwart, H. *Acquisition du langage et développement de la pensée: Sous-systèmes linguistiques et opérations concrètes*. Paris: Dunod, 1967.

Snow, C., & Ferguson, C. (eds.), *Talking to children: Language input and acquisition*. New York: Academic Press, 1977.

Stern, D. *The first relationship*. London: Fontana, 1978.

# Author Index

Page numbers in *italics* refer to the References

# Subject Index

H. Hörmann

# Psycholinguistics

An Introduction to Research and Theory
Translated form the German by H. H. Stern, P. Leppmann
2nd revised edition 1979. 60 figures. IX, 342 pages
ISBN 3-540-90417-4

„...provides a comprehensive introduction to the psychology of language by concentrating on the behaviourist
conception...
the translation is written in a clear, concise and compact
English...
The substance of this book, which has become a standard
textbook in German as well as the brilliancy of its translation will certainly secure its position in the English
speaking world as well.“  *IRAL (Deutschland)*

B. Malmberg

# Structural Linguistics and Human Communication

An Introduction into the Mechanism of Language and
the Methodology of Linguistics
(Reprint of the 2nd revised edition 1967)
1976. 88 figures. VIII, 213 pages
(Kommunikation und Kybernetik in Einzeldarstellungen,
Band 2)
ISBN 3-540-03888-4

**Contents:** Introduction. – Signs and Symbols. The
Linguistic Sign. – The Communication Process. – Preliminary Expression Analysis. Acoustic and Physiological
Variables. Information. – Segmentation. Forms of Expres-
sion. Oppositions and Distinctions. – Paradigmatic Struc-
tures. – Redundancy and Relevancy. Levels of Abstrac-
tion. – The Distinctive Feature Concept. The Binary
Choice. – Syntagmatic Structures. Distribution and Proba-
bility. – Content Analysis. – The Functions of Language. –
Perception and Linguistic Interpretation. – Primitive
Structures and Defective Language. – Linguistic Change. –
Bibliographical Notes. – Author Index. – Subject Index.

"A general survey of modern structural linguistics by
B. Malmberg...
The book is essentially intended for the advanced student,
but others will also find it useful, since the author manages
to deal lucidly and intelligibly with a difficult subject."
*The Years Work in English Studies*

Springer-Verlag
Berlin
Heidelberg
New York

M. Miller

# The Logic of Language Development in Early Childhood

Translated from the German by R. T. King
1979. 1 figure, 30 tables. XVI, 478 pages
(Springer Series in Language and Communication,
Volume 3)
ISBN 3-540-09606-X

**Contents:** Introduction. – Description of the Project:
Cognitive and Social Determinants of Language
Acquistion. – A Quantitative Analysis of the Early
Linguistic Development of Meike and Simone. –
Problems in the Transformational Analysis of Early
Child Language. – Aspects of the Early Linguistic
Development of Meike and Simone. – Conclusion. –
Appendix. – References. – Author Index. – Subjext
Index.

Springer-Verlag
Berlin
Heidelberg
New York

G. Hammarström

# Linguistic Units and Items

1976. 17 figures. IX, 131 pages
(Communication and Cybernetics, Volume 9)
ISBN 3-540-07241-1

**Contents:** Introduction. – Spoken Language. –
Written Language. – Written Language in Relation
to Spoken Language. – Spoken Language in Relation
to Written Language. – The Tasks for Linguistics. –
Bibliography. – Author Index. – Subject Index.

45
55
65

Classics